Exergy and Its Application - Toward Green Energy Production and Sustainable Environment

Edited by Muhammad Aziz

Published in London, United Kingdom

IntechOpen

Supporting open minds since 2005

Exergy and Its Application - Toward Green Energy Production and Sustainable Environment
http://dx.doi.org/10.5772/intechopen.77461
Edited by Muhammad Aziz

Contributors
Eduardo Torres-Sánchez, Xing Luo, Xu Zhu, Eng Gee Lim, Muhammad Aziz, Muhammad W. Ajiwibowo, Arif Darmawan, Mohammad Sardarabadi, Amin Farzanehnia, Dalibor Purkovic

First published in London, United Kingdom, 2019 by IntechOpen
IntechOpen is the global imprint of INTECHOPEN LIMITED, registered in England and Wales, registration number: 11086078, 7th floor, 10 Lower Thames Street, London, EC3R 6AF, United Kingdom
Printed in Croatia

British Library Cataloguing-in-Publication Data
A catalogue record for this book is available from the British Library

Additional hard and PDF copies can be obtained from orders@intechopen.com

Exergy and Its Application - Toward Green Energy Production and Sustainable Environment
Edited by Muhammad Aziz
p. cm.
Print ISBN 978-1-78984-674-4
Online ISBN 978-1-78984-675-1
eBook (PDF) ISBN 978-1-78985-353-7

For EU product safety concerns:
IN TECH d.o.o., Prolaz Marije Krucifikse Kozulić 3, 51000 Rijeka, Croatia, info@intechopen.com or visit our website at intechopen.com.

Meet the editor

Dr. Aziz is currently an associate professor at the Institute of Industrial Science, University of Tokyo, Tokyo, Japan. He received his BEng, MEng, and DEng degrees from Kyushu University, Japan, in 2004, 2006, and 2008, respectively. He previously worked for Seiko-Epson Corp., Japan, until 2009, and completed his postdoctoral at the University of Tokyo in 2011. He then moved to the Tokyo Institute of Technology, Japan, as assistant professor, after which he became an associate professor at the same university in 2015. His general research area is energy systems. His research interest includes power generation, renewable energy utilization, process modeling, smart grid, electric vehicles, batteries, and hydrogen production and utilization. He has authored more than 100 articles in peer-reviewed journals, and 16 books and book chapters. Recently, he received several awards including Outstanding Paper Award from the Journal of Chemical Engineering of Japan in 2013, Japan Institute of Energy Award for Encouragement in 2016, and the Best Paper Award from the Japan Society of Energy and Resources in 2018.

Contents

Preface

Exergy has been defined as the maximum work that is useful, extracted from any process toward its equilibrium. Hence, it has a very strong connection with the second law of thermodynamics. In energy harvesting and management systems, the concept of exergy is very important because it represents the efficiency of the system. Exergy can be used as a tool to measure resource efficiency, as well as whole system sustainability. In addition, it can also be used to analyze and clarify the performance of each process; hence, methods of improvement can be determined.

This book is the result of a very careful selection of chapters and contributors in the related field. The book is divided into three main sections according to the approaches and purpose of each proposed chapter. The first chapter in the "Introduction" section, deals with detailed exergy analysis of developed photovoltaic/thermal systems employing nanofluid, which are designed to harvest solar energy efficiently. Various models have been developed and compared with each other to achieve optimum design and conditions. In addition, the utilization of nanofluid as optical filters of solar radiation leads to higher exergy efficiency. The second section, "Advanced energy conversions," describes several advanced technologies that are considered to have great potential in energy conversion and harvesting, and comprises three chapters. The second chapter describes the idea of power-to-gas energy storage systems coupled with a combined cycle employing chemical looping combustion technology to facilitate base and intermediate loads of power utilizing unused or surplus electricity from the grid. In addition, the idea of chemical looping also results in very clean energy conversion due to its capability to separate the produced CO_2 during the reduction process. In this chapter, to achieve high total energy efficiency, the idea of enhanced process integration technology is employed resulting in efficient and optimal heat and exergy recovery throughout the system. The third chapter focuses on novel models and design for energy conversion (power generation) using the principles of electromagnetics. The text contains a detailed description related to electromagnetic induction, magnetic susceptibility, potential materials, and magnetic hysteresis. In addition, a diagram of electromagnetic systems is also provided and explained.

The third section focuses on the idea of "innovative energy management systems" toward high-quality energy systems and contains two chapters. The fourth chapter is dedicated to the utilization of electric vehicles for demand-side energy management. The vast deployment of electric vehicles is considered to have great potential when they can be utilized effectively for providing ancillary services to the grid. In addition, the utilization of electric vehicles in load scheduling can facilitate households to alleviate the network load burden, in addition to reducing electricity bills. Also in this chapter, two different models of demand response assisted using electric vehicles are described, including electric vehicle auxiliary power supply and neighbor energy-sharing models. The objective of the developed models is to optimize the load distribution for both individual and multihousehold networks via vehicle-to-home and vehicle-to-neighbor connections. Finally, the fifth chapter explains the utilization of supercapacitors as very responsive energy storage devices in low-power modules. The supercapacitor has many advantages compared to other types of energy storage, including high charging and discharging rates, very small capacity degradation, low toxicity of used materials, and high cycle efficiency. The chapter explains in detail the characteristics and performance of a lithium-ion supercapacitor, especially its usage for low-power sensor modules.

It is expected that this book will provide and enrich the state of the art in advanced energy systems, including energy conversion and management. All the chapters cover a broad range of disciplines, which are correlated in terms of the efforts toward efficient energy systems. In addition, the correlation between energy and exergy, and their understanding, are believed to be very important to improve energy efficiency and guarantee better energy quality.

Muhammad Aziz
The University of Tokyo,
Japan

Section 1

Introduction

Chapter 1

Introductory Chapter: Green Energy Systems

Muhammad Aziz

1. Introduction

As one of the very crucial elements of life, energy has given huge impacts in forming the domestic and international policies of the country, raising the environmental issues, changing the social dimensions, accelerating the economic growth, etc. The study related to energy has focused mainly on finding and establishing the new and efficient ways to produce (convert), store, transport, and utilize the energy. The conventional energy system, coupled with the huge demand of energy due to industrial activities, has resulted in various social, psychological, and environmental problems. Concretely, conventional combustion of fossil energy sources has accelerated the emission of greenhouse gases (GHGs), leading to high concern on climatic issues. In order to tackle down the above problems, several research have been developed, especially related to the way to achieve greener energy systems. Dincer and Zamfirescu [1] have proposed the concept of greenization, which refers to the efforts to greenize the energy systems, including the processes and applications with high sustainability.

Some academic efforts to clearly analyze the problems also include several theoretical approaches to correlate the theories to their real implications. The intensive correlation of thermodynamics toward sustainable energy systems and environment has been studied by Hammond [2]. In addition, Bejan [3] also used the theory of entropy generation minimization in order to analyze the energy policy. Discussing the energy policy, there is a strong relationship between the exergy destruction and environmental impact. Lower exergy destruction leads to higher energy efficiency of the system; hence, the systems consumes less energy input followed by lower environmental impact, leading to the sustainability of the system.

2. Challenges in energy systems

Figure 1 shows four essential pillars related to energy issues, covering security and resilience, economy, social, and environment. In energy security, recently, the issue is not only limited to the security of supply but also includes the capability of the system to respond and recover back when any disturbance in the energy system occurs. Therefore, the security and resilience correlate strongly to both quantitative and qualitative aspects of energy. Furthermore, the economic issues cover several major points including equity, participation, economic performance, and open opportunity to the whole players. All of the entities in the system can behave as energy producers and consumers simultaneously, leading to the concept of "prosumer." The equal opportunity provided to the all players involved in the system potentially accelerate and maximize the introduction of renewable energy, as well as improve the total economic performance. The environmental issues deal

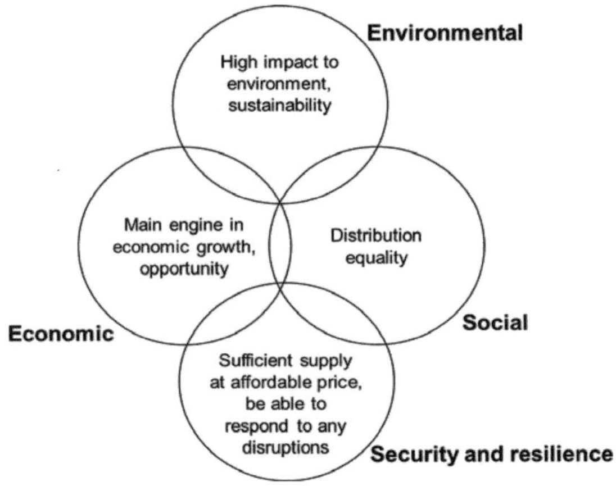

Figure 1.
Essential pillars in energy system.

strongly with some challenges to realize the green and clean energy system which give a minimum influence to the environment, as well as keep both the system and environment to be sustainable. At the last, the social issues are related strongly to the problems in equal and appropriate distribution of the energy to the consumers. Energy is very essential element in human activities; therefore, the correlation and balance of those four pillars in energy system are very important.

The transition of the energy systems from currently high- to low-carbon emitting systems has led to several socioeconomic-environmental opportunities. However, there are still many unresolved issues related to the correlation among the energy, economic growth, and social welfare. The prediction and projection of energy systems in the future, especially related to the balance of both fluctuating demand and supply, face many uncertainties, due to several parameters including social, economic, weather, climate, and technological developments.

Long-term accurate and flexible (responsive) strategies on energy will gave a strong confidence to all of the energy entities involved in the system/region. The dynamic and fast changes in energy systems have occurred time by time following the technological and social changes in the community.

3. Exergy

Energy has two different measures, which are quantity and quality. The energy quantity is the amount of energy which is supplied or consumed, while the energy quality deals with the availability/usability of energy resource to be consumed. The latter has revealed the concept of exergy. It is very strongly correlated to both the first and second laws of thermodynamics. The exergy is possibly destructed, which is different with the concept of energy as it is conserved. It strongly relates to the concept of entropy generation; however, exergy can be useful in determining the system components which are responsible for the system irreversibility [4].

The concept of exergy is very useful for economic, ecological, and sustainability measures, as its capability to describe the quality of the energy resource. The exergy analysis also includes and considers the impact of energy stream to the environment, which further influences significantly both energy and environmental policies. The exergy is a very potential indicator for the policy makers and industrial practitioners in justifying the energy system and resources [5].

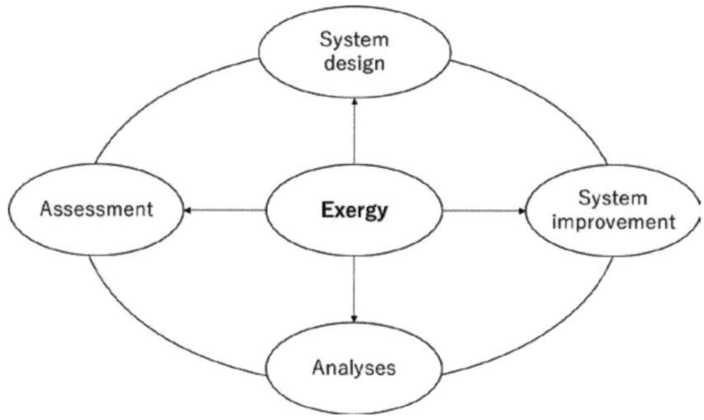

Figure 2.
The exergy concept for practical applications.

Figure 2 shows the possible utilization of exergy in practical applications. First of all, the exergy is very important to model and optimize the developed design of the processes or systems, before they are being developed and constructed. Furthermore, exergy is also convenient tool to analyze the performance of the developed system, as well as measure and predict the possible improvement points which can be performed.

The exergy of the system, *Ex*, is basically represented as follows:

$$Ex = S(T - T_0) - V(p - p_0) + \sum_i n_i(\mu_i - \mu_{io}) \tag{1}$$

where S, T, V, and p are entropy, temperature, volume, and pressure, respectively. In addition, n_i and μ_i are number of moles of substance i and chemical potential of substance i. From Eq. (1), it can be understood that the exergy could be zero when the system reaches the equilibrium with the surrounding environment. Practically, the exergy depends on several condition parameters, including temperature, pressure, chemical potential, electricity, magnetism, gravity, and radiation [6].

By adopting the exergy, both the quantity and quality of energy are coherently considered, resulting in the holistic approach of the energy itself. This is compatible for all cases of energy systems and all types of energy resources. Therefore, the economic and social prices of the energy systems and resources can be more clearly defined, as well as their impacts to the environment.

4. Conclusion

Development of sustainable energy system requires comprehensive knowledge, deep scenario analysis, and referable case studies. The discussion of energy requires a deep analysis of both quantitative and qualitative aspects. In addition, energy sustainability strongly imposes sustainability in other sectors, including economic, social, and environmental. The concept of exergy potentially leads to the opportunity for energy efficiency in the targeted system. The coherent understanding and application of both energy and exergy is urgently demanded. The inclusion of exergetic consideration in energy policy and practical application is strongly expected to be able to realize the sustainable and efficient energy system in the future.

Author details

Muhammad Aziz
Institute of Innovative Research, Tokyo Institute of Technology, Tokyo, Japan

*Address all correspondence to: maziz@ssr.titech.ac.jp

References

[1] Dincer I, Zamfirescu C. Potential options to greenize energy systems. Energy. 2012;**46**:5-15

[2] Hammond G. Engineering sustainability: Thermodynamics, energy systems, and the environment. International Journal of Energy Research. 2004;**28**:613-639

[3] Bejan M. Energy Policy, in Entropy Generation through Heat and Fluid Flow. New York: Wiley; 1994

[4] Jahangiri P, Sangi R, Thamm A, Streblow R, Müller D. Dynamic exergy analysis – part II: A case study of CHP district heating in Bottrop, Germany. In: Building Simulation and Optimization Conference (BSO14), 23-24 June 2014. London: UCL; 2014

[5] Favrat D, Marechal F, Epelly O. The challenge of introducing an exergy indicator in a local law on energy. Energy. 2008;**33**:130-136

[6] Wall G. Exergy flows in industrial processes. Energy. 1988;**13**:197-208

Section 2

Advanced Energy Conversion

Chapter 2

Chemical Looping Combustion Power Generation System for a Power-to-Gas Scheme

Muhammad W. Ajiwibowo, Arif Darmawan and Muhammad Aziz

Abstract

Renewable energy provides a quick win solution for global warming, but it comes with drawbacks. Renewable sources such as solar and wind are not available for continuous use; thus, intermittency of electric power generation is an issue. Fluctuation of electricity production could damage the grid. Throughout the years, researchers have come up with solutions to solve this problem by storing the excess electricity via an energy storage system. One of the most efficient options is through solid oxide electrolysis cell (SOEC) to produce H_2. In itself, H_2 contains a lot of energy and can be converted to electricity via combustion or fuel cell. Therefore, storing electricity in the form of H_2 could prove to be effective. Energy storage systems such as power-to-gas may provide a clean and efficient way to store the overproduced electricity. In this work, a power-to-gas system coupled with a chemical looping combustion combined-cycle system is proposed to provide base and intermediate load power from the unused electricity from the grid. Enhanced process integration was employed to achieve optimal heat and exergy recovery. This chapter focuses on the design of a system consisting of a power-to-gas conversion method and a H_2-powered chemical looping combustion power generation system.

Keywords: chemical looping combustion, solid oxide electrolysis cell, system modeling, power to gas

1. Introduction

Combustion of fossil energy sources for industrial processes around the world contributes massively to the creation of greenhouse gases (GHG) and mainly CO_2. This has been a major problem worldwide as it directly increases the pollution level and the likelihood of the earth's increase in temperature. This led the international community to investigate ways to prevent this phenomenon [1]. Besides, fossil fuels are inevitably bound to be depleted in the future [2]. As of right now, as much as 84% of the world energy consumption is still fossil fuel driven, less than what the previous year had (85%); even so, it is still very high. Without a radical transformation, fossil fuel will still be the majority of energy source in the foreseeable future. This condition necessitates scientists and engineers to provide a sound solution. Efforts to reduce the fossil fuel usage and transition to sustainable energy

 IntechOpen

sources remain a challenge for scientists and engineers alike. Alternative renewable and clean energy technologies are actively being developed now [3, 4]. Fossil fuel combustion for power generation and industrial processes around the world is a key contributor to CO_2 emissions and had caused the earth's temperature to increase ever since the industrial revolution. This brought the international community to implement emission regulations and policies to mitigate GHG effect on the earth while transitioning to sustainable alternatives to fossil fuels [1]. This is also supported by the fact that fossil fuel reserves are depleting due to massive use around the world [2]. Right now, as much as 84% of the world energy consumption is still derived from fossil fuels, which is less than what the previous year had (85%). Without a radical change, fossil fuel will still be the key energy source in the foreseeable future. Thus, efforts to reduce the fossil fuel usage and transition to sustainable energy sources remain a big challenge. Alternative renewable and clean energy technologies are actively being developed now [3, 4].

It is widely agreed and practiced that the conversion of fossil fuels into energy is mainly through combustion. It is an extremely efficient process and very mature in terms of technology. Its development dates back to the seventeenth century when the first steam engine was introduced. Regardless of its convenience now, combustion processes release lots of GHG and harmful gases (e.g., CO_2, SO_x, NO_x). Even so, combustion technologies will still thrive as a primary contributor to the world's power generation. This necessitates scientists to develop a combustion method that is also environmentally friendly. Many research and development efforts have been put into finding new combustion methods. One of the major drawbacks of traditional method for combustion utilizes air, where, upon reaction, NO_x and sometimes SO_x will form and will potentially cause health indications if inhaled over time. One of the most promising solutions for a clean combustion process is to react the fuel directly with pure oxygen, or simply called oxy-combustion. This method will only generate CO_2 and H_2O emissions. Although being clean and efficient, the drawback of this process is that it requires prior separation of oxygen from the air in the atmosphere, which directly increases cost and efficiency penalty for the power plant.

Moreover, one similar solution, the emission could just be removed completely if renewable energy is used. Even so, the drawback of renewable energy usage is the intermittency of its production. Thus, the development of renewable energy systems throughout the years has generated a new interest in energy storage technologies. It is a dominant contributor in a renewable energy system. Despite being sustainable and clean, it has been reported that the use of renewable energies from solar and wind power sources has caused burdens to the electricity grid. Energy storage technologies include H_2, batteries, flywheels, compressed air, ultracapacitors, pumped hydro, and compressed gas [5]. Energy storage could mitigate power variations, enhance the system's flexibility, and provide a scheme where surplus electricity from the grid could be stored and dispatched on demand [6].

Driven by the same problem, overproduction of electricity is also being demonstrated in Indonesia where renewable energy is not widely available yet. Heavily reliant on fossil fuels, the country consumes as much as 87.5% of all its power generation right now [7, 8]. Furthermore, according to the National Electricity Company in Indonesia (PLN), by default, Java-Bali sector of Indonesia's power generation generates as much as 6000 MWe excess electricity production, which is wasted due to lack of energy demand. Likewise, this indicates a strong need for energy storage option. It is widely considered that the conversion of fossil fuels into energy is mainly through combustion. The development of combustion processes for power dates back to the seventeenth century when the steam engine was first founded. With all its convenient characteristics, combustion process releases lots of GHG and harmful gases (e.g., CO_2, SO_x, NO_x). Seeing that combustion technologies will remain as a primary contributor to the world's power generation, many research and developments have been put into finding new combustion methods. One of the promising ideas for clean combustion is to react the fuel directly with pure oxygen, or simply called oxy-combustion. This method will only generate CO_2 and H_2O emissions. Although being clean and efficient, this method requires prior separation of oxygen from the atmospheric air, which may increase cost and cause efficiency penalty for the power plant.

Moreover, the GHG emission could just be removed completely if renewable energy is used. The development of renewable energy systems throughout the years have generated a new interest in energy storage technologies. It is a dominant contributor in a renewable energy system. Despite being renewable and clean, it has been reported that the use of renewable energies from solar and wind power sources have caused burdens to the electricity grid. Energy storage technologies include H_2, batteries, flywheels, compressed air, ultracapacitors, pumped hydro, and compressed gas [5]. Energy storage could mitigate power fluctuations, enhances the system's flexibility and provides a scheme where surplus electricity from the grid could be stored and dispatched on demand [6].

By the same token, overproduction of electricity is also being demonstrated in Indonesia, where renewable energy is not widely available yet. Heavily reliant on fossil fuels, the country consumes as much as 87.5% of all its power generation right now [7, 8]. Furthermore, according to the stated-owned electricity company in Indonesia, PLN, by default, Java-Bali sector of Indonesia's power generation generates as much as 6000 MWe excess of electricity production, which are wasted due to lack of energy demand. Likewise, this indicates a strong need for energy storage option.

1.1 Hydrogen production and utilization

Hydrogen as an alternative energy source is predicted to have a powerful role in the low-carbon future [9]. It is long recognized as a sustainable fuel due to its favorable characteristics [10]. Besides that, hydrogen is greatly abundant on earth, albeit in its oxidized state (H_2O) [11]. Presently, H_2 is commercially produced using reforming technologies on hydrocarbon fuels. Researches for the production of H_2 have been a great interest for a long time. The routes to produce H_2 vary from chemically to electrochemically. Chemically, supercritical water gasification (SCWG) of biomass and syngas chemical looping (SCL) are among the most efficient ways to produce hydrogen [11, 12]. Compared to other types of gasification, SCWG utilizes more steam, thus promoting more reforming of the gases to produce H_2.

Moreover, from the electrochemical routes, many types of water electrolysis are considered to be effective for H_2 production by splitting water. Proton-exchange membrane (PEM) electrolysis and alkaline water electrolysis are the most used technology due to their maturity [13]. Typically, PEM electrolysis and alkaline water electrolysis have energy efficiencies of around 50–80%. Furthermore, solid oxide electrolysis cell (SOEC) is another type of electrolysis method that utilizes a high temperature of around 750–1000°C and 10–15 bar of pressure [14]. In higher temperatures, higher efficiency could be achieved due to smaller energy requirements.

The utilization of H_2 also varies from just energy storage to electric power production. Various H_2 energy conversion technologies are already generally understood, and many are already under commercial real-world developments

[10]. Fuel cells are among the best and are among the most efficient ways for energy conversion from H_2 for electric power production [15]. Integration of fuel cells with other technologies remains a challenging puzzle. Various integrations of fuel cell and other systems are currently being investigated [16]. Generally, fuel cell is very versatile in terms of fuel types and various operating conditions. Nevertheless, one of the major drawbacks of fuel cell is that the membrane used in the cells may have a relatively short lifetime and, hence, could potentially add costs to fuel cell systems.

Numerous efforts have been made to integrate fuel cells, especially SOEC with other energy conversion technologies. Cinti et al. proposed and investigated an integrated SOEC and Fischer-Tropsch system to produce methane from surplus renewable energy [17]. Energy and exergy evaluation of an SOEC-methanation system is also evaluated by Luo et al. [18]. Kezibri et al. also modeled a power-to-gas system for oxy-combustion power generation [19]. But they utilized and considered a proton-exchange membrane-based electrolyzer, which is heavily reliant on the membrane and did not further evaluate the possibilities of system and heat integration. All these efforts can prove high potential of hydrogen energy utilization in the future. Unfortunately, there is still no significant effort made to provide an efficient energy system utilizing SOEC technology for hydrogen and, subsequently, power.

1.2 Chemical looping combustion power system

Chemical looping combustion (CLC) is a new and leading-edge energy conversion method that utilizes metal oxides or otherwise called oxygen carriers to oxidize the fuel instead of atmospheric air. It uses two reactors, namely, the reducer and the oxidizer. Mixtures of metal oxide and some inert solids for heat dilution are being circulated throughout these two reactors. The process starts when fuel is oxidized in the reducer reactor; afterwards, the reduced metal oxide will be re-oxidized by air in the oxidizer while also generating heat in the process. These processes are being repeated in a looping fashion. There have been many reports on the use of various types of fuels, including hydrocarbon and biomass. Typically, exhaust gas containing CO_2 and H_2O will be generated from the fuel oxidation, and CO_2 will be easily separated by condensation. Inherent CO_2 separation and storage are possible in this system due to the separate combustion processes. Furthermore, if H_2 is used, the combustion product will be just H_2O. Thus, this combustion technology has a high potential as a zero emission power generation system due to its favorable characteristics, namely, the inherent separation capability for CO_2. As a comparison, this system could achieve the same advantage that an oxy-combustion power plant has without the need to separate air first with an air separator unit (ASU).

Thus, this paper tries to focus on the effort to propose an efficient energy system which comprises of SOEC and chemical looping combustion (CLC) combined cycle power generation to produce baseload power from hydrogen as a sustainable zero emission energy source. The produced H_2 is fed to the CLC power generation system where it is mixed with natural gas. This way, environmental burden caused by GHG can be decreased as we are producing less CO_2 compared to traditional natural gas-fired power generation system. Detailed process integration and the calculation are discussed in Sections 2 and 3. Besides that, parameters and operation conditions are evaluated to further improve the system's energy conversion efficiency and its key design parameters.

This paper focuses on the effort to propose an efficient energy system that comprises of SOEC and chemical looping combustion (CLC) combined cycle power generation to produce base and intermediate loads power from both natural gas and renewable energy-based hydrogen. The produced H_2 from renewable energy source is fed to the CLC power generation system where it is mixed with natural gas. By this way, environmental burden caused by GHG can be decreased as the process is producing less CO_2 compared to traditional natural gas fired power generation system. In addition, the inherent fluctuation characteristic of renewable energy can be stabilized, producing a stable and reliable power generation to the grid. Detailed process integration and the calculation is discussed in Sections 2 and 3. Besides that, parameters and operation conditions are evaluated to further improve the system's energy conversion efficiency and its key design parameters.

2. Proposed integrated system

2.1 Integrated SOEC and chemical looping combustion power generation system

Figure 1 depicts a simplified process flow diagram for the proposed system. The system itself consists of an SOEC as the H_2 producer module and a chemical looping combustion combined cycle (CLCCC) as an electric power generation module. Theoretically, the CLCCC could act as the main electric power generator providing a stable electrical power output that is suitable for base and intermediate loads. The CO_2 emitted from the combustion process is basically separated. This leads to low CO_2 emission and potentially zero CO_2 emission if H_2 is used. For this system, a dual fuel system is considered where natural gas is considered, while the H_2 produced from the electricity from renewable energy sources or surplus electricity from the grid becomes additional fuel. Theoretically, the produced H_2 is consumed without being stored; therefore, the flow of natural gas to CLCCC decreases accordingly. Although, a storage infrastructure could also be considered for H_2 in such system, it is not considered for this study. Hence, the generated electric power from the system is assumed stable, without being influenced by the fluctuation that comes with renewable energy sources and grid surplus electricity.

Overall, surplus electricity from the grid is used to convert H_2O into H_2 and O_2 via the SOEC process. Afterwards, as described in the previous part, the H_2 is directed and fed directly to the CLCCC module. On the other hand, additional O_2 is also being fed into the CLCCC power system along with air, which potentially leads to higher combustion temperature. **Table 1** describes the main assumptions and parameters used for the developed CLCCC system.

In order to achieve highest energy efficiency for the system, enhanced process integration methodology is utilized. This approach primarily focuses on heat and exergy recovery in the system via heat exchanger integration and compression [20, 21].

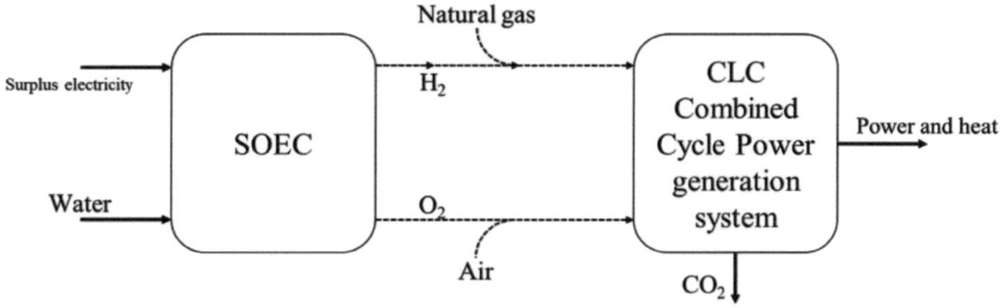

Figure 1.
General overview of the integrated system.

Component/system	Unit	Value
Solid composition		70% metal oxide
		15% SiC
		15% Al_2O_3
Generator efficiency	%	98
Compressor isentropic efficiency	%	90
Turbine isentropic efficiency	%	90
Fuel flow rate	kg/s	6
OT inlet temperature	°C	1400
RT inlet temperature	°C	800–900
ST inlet temperature	°C	700
Operating pressure	Bar	15–35
ST inlet pressure	Bar	250

Table 1.
Details on the parameters and assumptions used in the CLCCC.

Waste heat from hot downstream processes is utilized to support the heat requirements of upstream processes recuperatively. Exergy is also elevated in cold streams via compression. It is a proven methodology that has been demonstrated by various works for producing electricity or hydrogen from various sustainable sources, especially biomass [9, 11, 22].

Basically, the CLCCC is somewhat similar to a traditional combined cycle with a gas turbine, but the combustor in such system is replaced by two chemical looping combustion reactors that act as the heat source for the downstream turbines. As opposed to the traditional method, the produced fuel gas that is rich in CO_2 could be directly separated as it does not produce other by-products except CO_2 and H_2O. Thus, the separation of CO_2 is significantly less energy intensive than the traditional separation method. The schematic diagram of CLCCC is shown in **Figure 2**. In a dual fuel scenario, H_2 and CH_4 (natural gas) are considered to be fuels for this system, with the key assumption of a reactor design that would support the use of these two fuels. Two fuel gas streams coming out of the CLC process, namely, the reducer gas and the oxidizer gas, are expanded via the reducer turbine (RT) and the oxidizer turbine (OT). Afterwards, the CO_2-rich stream leaving the RT is directly separated by condensation and then compressed and stored. On

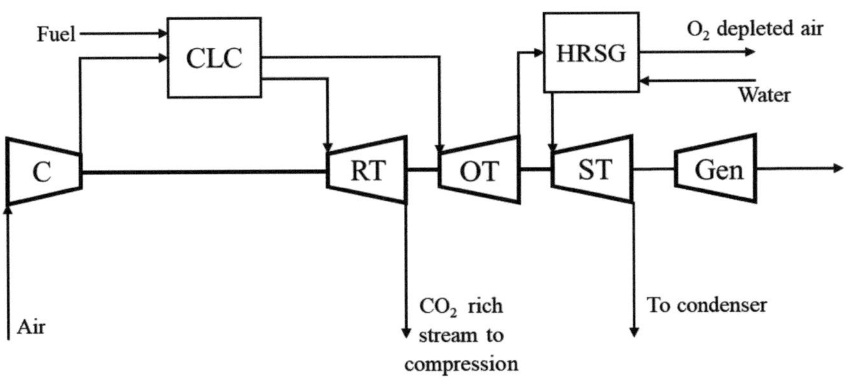

Figure 2.
CLCCC power generation system.

Component/system	Unit	Value
Power input	MW	350
Produced hydrogen	Nm^3/h	87,803
Produced oxygen	Nm^3/h	43,901
Operating temperature	°C	750
Operating pressure	Bar	10
Efficiency	%	88

Table 2.
Assumptions for the SOEC used in the model.

the other hand, the high-temperature gas leaving the OT is used to generate steam for generating more power via a steam cycle.

2.2 Solid oxide electrolysis cell for H_2 production

As discussed in the previous section, SOEC is regarded as one of the most efficient electrolysis processes to produce pure H_2 [23]. For the purpose of this study, SOEC is considered favorable if it is used for power-to-gas energy storage. In this case, the input electricity used for the SOEC is the surplus electricity from the grid (especially due to surplus from renewable energies) or directly from renewable energy sources. Afterwards, the generated H_2 will be stored temporarily or can be directly integrated with the CLCCC system described before. The SOEC parameters in this study are based on the research conducted by Udagawa et al. [23]. The detailed parameters are provided in **Table 2**.

3. Process modeling and calculation

For the purpose of system, mass, and energy balance simulation, ASPEN Plus V8.8 from Aspen Technology, Inc., is utilized in this study. Key assumptions made for this model are listed in **Table 1** that are primarily taken from other experimental and numerical researches. The operating conditions are chosen based on other literatures [23–25]. Key thermodynamic assumptions are as follows: (i) ambient temperature is set to 27 °C; (ii) no heat loss is assumed; and (iii) air is assumed to contain 79% mol N_2 and 21% mol O_2. RStoic reactor blocks are used to model the reducer and oxidizer reactors in ASPEN Plus. Simplistically, in the reducer, the metal oxides will be reduced by the fuels, which are H_2 and CH_4, and then circulated to the oxidizer where it is re-oxidized by air. The operating reactors are assumed to be an entrained flow type for the oxidizer, and a moving bed type is used for the reducer.

To identify different parameters used in this study, three types of metal oxides are evaluated for the CLCCC process. Two of the most studied oxygen carriers, nickel oxide (NiO) and iron oxide (Fe_2O_3), are each utilized for the CLCCC process. Additionally, $CaSO_4$, also known as gypsum, is also utilized and evaluated in this study due to recent interests for this material as an oxygen carrier due to its favorable chemical characteristics [26]. All of these metal oxides have many distinctive characteristics, such as resistance to elevated temperatures, high oxygen concentration, and so on that could provide extra benefit and efficiency boost to the CLCCC system. **Table 2** provides the description of the metal oxides used in this study.

And then, for parametric purpose, different H_2 and CH_2 mix flow rates are considered. It is aimed to simulate different intermittent or fluctuating renewable energy source used to generate H_2 in the SOEC. The key assumption is that the heat rate for the CLCCC is assumed to be the same for each different mixture. The base case is 6 kg/s of H_2. And then, for the metal oxide mixture, inert solids are also considered as heat diluents and heat carriers. The mass fraction is assumed to be 70% metal oxide and 30% inert materials, which consist of 15% SiC and 15% Al_2O_3 as suggested by Fan [25]. **Figure 3** shows the detailed process flow diagram of the proposed system (**Table 3**).

As described in previous parts, the process begins with the fuel produced in the SOEC entering the CLCCC along with natural gas. The mixture is then directly combusted in the reducer. The fuel gases leaving the reducer and oxidizer are expanded in the RT and OT, respectively, for electricity production. The CO_2 leaving RT is separated and compressed up to 200 bars for storage. Besides that, the fuel gas leaving the OT is utilized for heat addition of air and water streams. On the other hand, the heat in reducer gas is also used to heat the water feed stream in HX-1 and compressed air in HX-2. And then, HX-2 is also utilized to cool down the compressed air coming from C2 to reduce additional compression work in downstream processes.

The system's total energy efficiency is defined and calculated as follows:

$$W_{Gen} = W_{RT} + W_{OT} + W_{ST} \tag{1}$$

$$W_{Used} = W_{Compressors} + W_{Pump} + W_{Auxilary} \tag{2}$$

$$\eta_{tot} = \frac{\text{W}_{\text{gen}}\text{-W}_{\text{used}}}{\text{LHV}.\dot{\text{m}}_{\text{fuel}}} \tag{3}$$

where, W_{GEN}, W_{RT}, W_{OT}, and W_{ST} are the total work obtained from the system, works obtained from RT, OT, and ST, respectively. In addition, W_{Used},

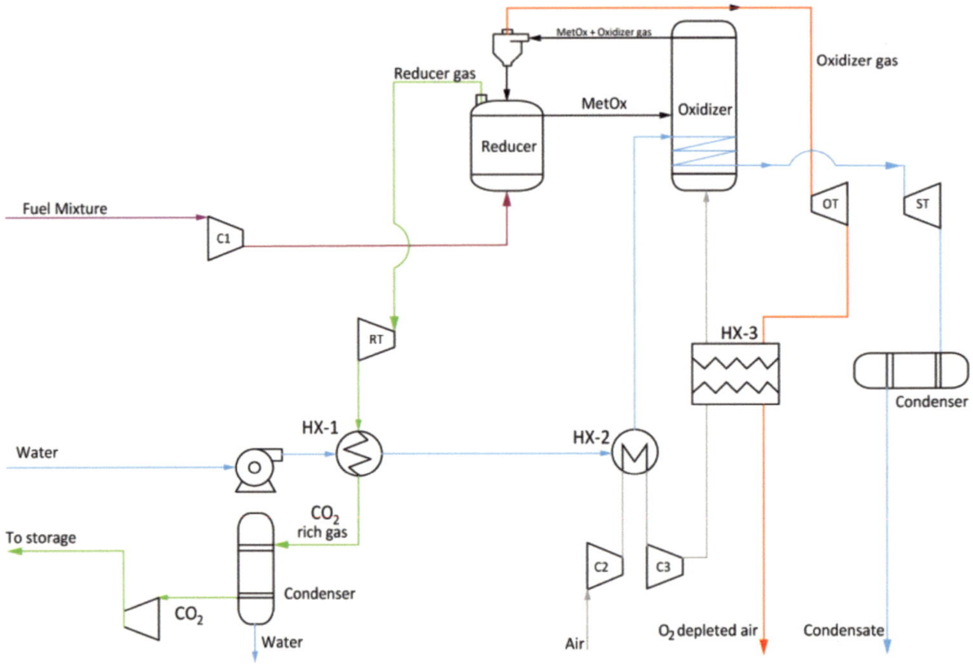

Figure 3.
Detailed process flow diagram of the CLCCC power generation process.

Metal oxide	Characteristic		
	Minimum melting point (°C)	Oxygen transport capacity (wt%)	Representative reactions*
Fe_2O_3/FeO	1565	10	(R)$Fe_2O_3 + H_2 \rightarrow 2FeO + H_2O$
			(R)$FeO + H_2 \rightarrow Fe + H_2O$
			(R)$4Fe_2O_3 + CH_4 \rightarrow 8Fe + 3CO_2 + 6H_2O$
			(O)$4Fe + 3O_2 \rightarrow 2Fe_2O_3$
NiO/Ni	1455	21.4	(R)$NiO + H_2 \rightarrow Ni + H_2O$
			(R)$4NiO + CH_4 \rightarrow 4Ni + CO_2 + 2H_2O$
			(O)$2Ni + O_2 \rightarrow 2NiO$
$CaSO_4/CaS$	1460	47.06	(R)$CaSO_4 + 4H_2 \rightarrow CaS + 4H_2O$
			(R)$CaSO_4 + CH_4 \rightarrow CaS + CO_2 + 2H_2O$
			(O)$CaS + O_2 \rightarrow CaSO_4$

**(R) reactions occur in the reducer and (O) reactions occur in the oxidizer.*

Table 3.
Details of the oxygen carriers used in the system.

$W_{Compression}$, W_{Pump}, and $W_{Auxiliary}$ are the total consumed work by the system, works consumed by compressors, pumps, and auxiliaries, respectively. Finally, the total energy efficiency η_{tot} is defined as the ratio of net produce power to the total calorific value of the fuels, including H_2 and CH_4.

4. Results and discussion

The parametric analysis has been done to provide a better understanding of the key design parameters of the SOEC-CLCCC system. First of all, as mentioned in the previous section, $CaSO_4$, Fe_2O_3, and NiO are each adopted and evaluated in the CLCCC system as oxygen carrier. Afterwards, different operating pressures from 15 to 35 bars, with an interval of 5 bars, are simulated for each system. Furthermore, different percentages for the H_2 used as fuel are also being investigated.

For the case of $CaSO_4$ as the oxygen carrier, the results are presented in **Figure 4**. The effects of different pressures and fuel mixtures are evaluated in the graph. The main characteristics of this particular oxygen carrier are high oxygen transport capacity, high temperature resistance, and highly exothermic reactions with the fuels used. Generally, the main driver for efficiency is the different fuel mixture. The highest energy efficiency obtained is 53%, with full H_2 feed. The efficiency decreases along with decreasing H_2 use. Basically, more CH_4 is required with each incremental reduction of H_2. Thus, more compression duty is required for the fuels due to higher flow rate. By the same token, lower pressures are more favorable for this case due to high compression work required. Thus, 15 bars achieve the highest efficiency of 54%.

Afterwards, when Fe_2O_3 is used as oxygen carrier, the highest energy efficiency obtained by the system rises to about 56%. The relationship is depicted in **Figure 5**. Basically, the trend is similar to the $CaSO_4$ system. When the H_2 percentage decreases, the energy efficiency decreases accordingly to about 45–47%. And then, lower pressures tend to be more preferable to achieve higher energy efficiency. This

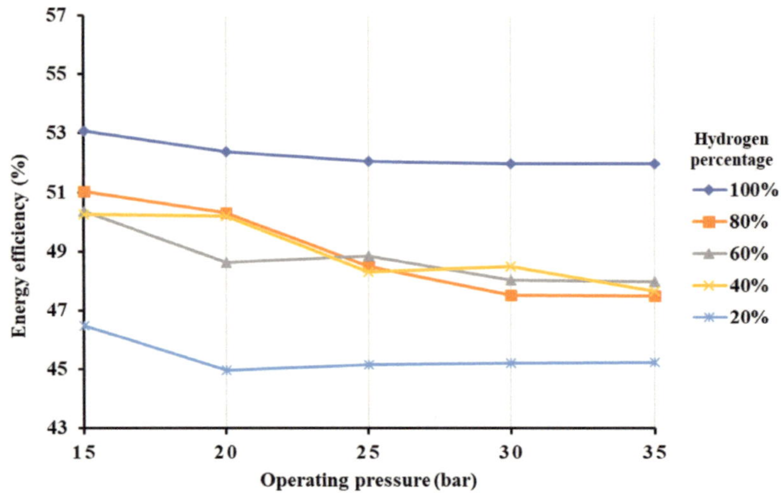

Figure 4.
System energy efficiency vs. operating pressure for the $CaSO_4$ system.

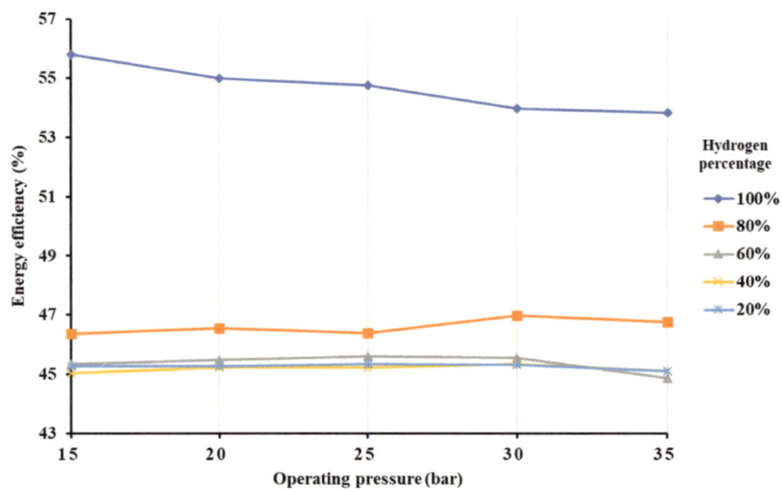

Figure 5.
System energy efficiency vs. operating pressure for the Fe_2O_3 system.

oxygen carrier has the lowest oxygen carry capacity compared to the other two oxygen carriers, but it has the highest melting temperature.

Furthermore, **Figure 6** shows the relationship of the operating pressures and the hydrogen percentages to the energy efficiency when NiO is used as oxygen carrier. In this case, the achievable maximum energy efficiency is about 56%. When H_2 percentage is reduced, the energy efficiency of the system further decreases to about 45–47%. Moreover, when larger amount of CH_4 is supplied, larger energy is required for CO_2 compression. NiO as an oxygen carrier has the second highest oxygen carry capacity and melting temperature compared to the other two oxygen carriers used in this study. These differences in efficiencies are driven by the difference of the heat released by the reactions that exist in the reducer and oxidizer.

Generally, from process modeling, operating pressure of the CLC system plays a significant role on changing the system's power intake and production. Process modeling suggests that the highest compressor work is consumed for the air compression process. Due to lower heating value of CH_4 compared to H_2, the system

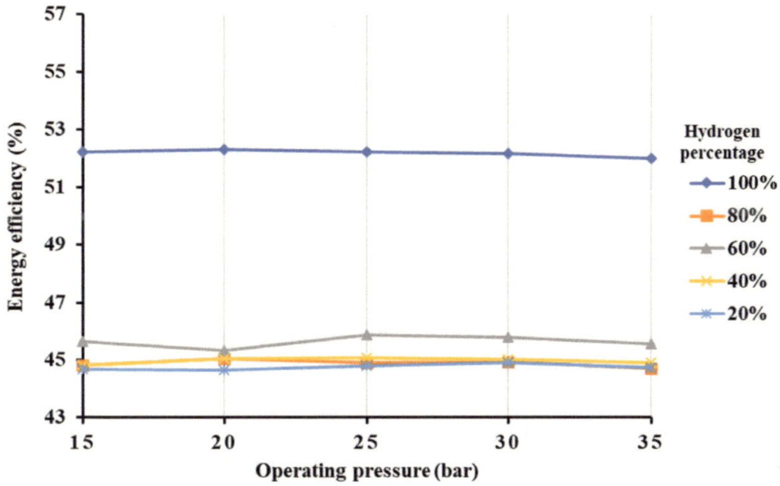

Figure 6.
System energy efficiency vs. operating pressure for the NiO system.

energy efficiency decreases accordingly when larger amount of CH_4 is used as fuel. Besides that, the thermodynamic characteristics also played a significant role to determine the heat produced and requirement in the combustion system. Subsequently, in case of $CaSO_4$ is used as oxygen carrier, as the reaction of CH_4 with $CaSO_4$ is endothermic, the input energy is required. This is opposite to the H_2 that provides an exothermic reaction releasing a considerable amount of heat. From process modeling, it can be inferred that the highest amount of heat exchange occurs in the oxidizer, where the heat from the oxidizer is utilized to heat up the steam to a higher temperature.

Compared to other power generation systems, the proposed system does not require additional process for air separation unit (ASU) and CO_2 separation process. Yet, the proposed system can achieve relatively high energy efficiency, which is similar to the energy efficiency of oxy-combustion power system that requires further CO_2 separation process and ASU.

Figure 7 shows the required solid flow rate for each different system. It shows that $CaSO_4$ requires the least amount of solids for the system for complete reduction of fuel and also for heat dilution. $CaSO_4$ has the highest O_2 transport capacity compared to the other two oxygen carriers. Fe_2O_3 comes second and NiO comes third in this comparison. The use of solids for heat dilution has to be considered for

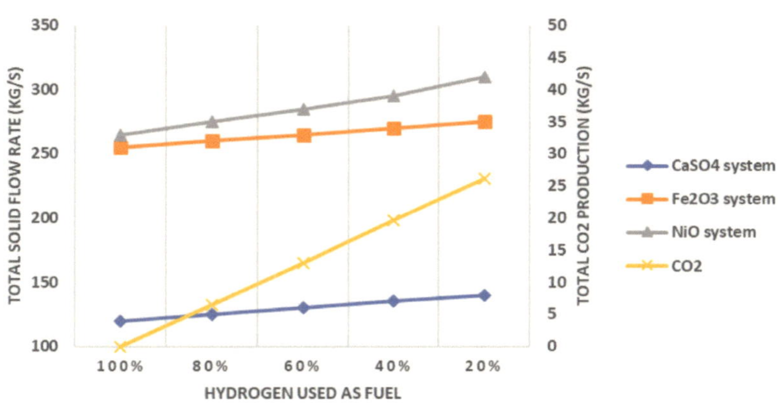

Figure 7.
Total solid flow rate and the CO_2 production for each system.

the CLCCC system, because air that is usually used for heat dilution could be replaced totally with solids. For further optimization of the system, excess air or excess solids can be considered as temperature reduction agents in the oxidizer. Higher amounts of solids circulated will require bigger reactors and more solid control infrastructure, and higher amount of air flow would require more compressor work.

5. Conclusion

A clean and efficient energy system to utilize efficiently hydrogen produced from renewable energy or surplus electricity is proposed. The proposed system is based on the chemical looping combustion using CH_4 as the base fuel. The system consists of SOEC for hydrogen production, chemical looping combustion, and combined cycle for power generation. The cleanliness of the developed system is promising as it can separate CO_2 directly via condensation due to the clean combustion process of CLC. A high energy efficiency of about 56% can be obtained. The results of this study are useful for further improvement and developments toward an actual CLC power generation system.

Author details

Muhammad W. Ajiwibowo[1*], Arif Darmawan[2] and Muhammad Aziz[3]

1 Department of Mechanical Engineering, Universitas Indonesia, Depok, Jawa Barat, Indonesia

2 Department of Transdisciplinary Science and Engineering, Tokyo Institute of Technology, Yokohama, Kanagawa, Japan

3 Institute of Innovative Research, Tokyo Institute of Technology, Tokyo, Japan

*Address all correspondence to: mwajiwibowo@gmail.com

References

[1] Szulejko JE, Kumar P, Deep A, Kim KH. Global warming projections to 2100 using simple CO_2 greenhouse gas modeling and comments on CO_2 climate sensitivity factor. Atmospheric Pollution Research. 2017;**8**(1):136-140

[2] Sorrell S, Speirs J, Bentley R, Brandt A, Miller R. Global oil depletion: A review of the evidence. Energy Policy. 2010;**38**(9):5290-5295

[3] Nasruddin N et al. Potential of geothermal energy for electricity generation in Indonesia: A review. Renewable and Sustainable Energy Reviews. 2016;**53**:733-740

[4] Surjosatyo A, Haq I, Dafiqurrohman H, Gibran FR. Effect of rice husk ash mass on sustainability pyrolysis zone of fixed bed downdraft gasifier with capacity of 10 kg/hour. AIP Conference Proceedings. 2017;**1826**:020009

[5] Ould Amrouche S, Rekioua D, Rekioua T, Bacha S. Overview of energy storage in renewable energy systems. International Journal of Hydrogen Energy. 2016;**41**(45):20914-20927

[6] Wendel CH, Kazempoor P, Braun RJ. Novel electrical energy storage system based on reversible solid oxide cells: System design and operating conditions. Journal of Power Sources. 2015;**276**:133-144

[7] PwC. Power in Indonesia: Investment and Taxation Guide. Vol. 9. PwC Publ; 2017. p. 132

[8] Lasnawatin F, Indarwati F. Handbook of Energy and Economic Statistics of Indonesia 2017. Ministry of Energy and Mineral Resources Republic of Indonesia; 2018

[9] Zaini IN, Nurdiawati A, Aziz M. Cogeneration of power and H_2 by steam gasification and syngas chemical looping of macroalgae. Applied Energy. 2017; **207**:134-145

[10] Nikolaidis P, Poullikkas A. A comparative overview of hydrogen production processes. Renewable and Sustaiable Energy Reviews. 2017;**67**: 597-611

[11] Aziz M. Combined supercritical water gasification of algae and hydrogenation for hydrogen production and storage. Energy Procedia. 2017;**119**: 530-535

[12] Darmawan A, Ajiwibowo MW, Yoshikawa K, Aziz M, Tokimatsu K. Energy-efficient recovery of black liquor through gasification and syngas chemical looping. Applied Energy. 2018; **219**:290-298

[13] Chi J, Yu H. Water electrolysis based on renewable energy for hydrogen production. Chinese Journal of Catalysis. 2018;**39**(3):390-394

[14] Jensen SH, Sun X, Ebbesen SD, Knibbe R, Mogensen M. Hydrogen and synthetic fuel production using pressurized solid oxide electrolysis cells. International Journal of Hydrogen Energy. 2010;**35**(18):9544-9549

[15] Gómez SY, Hotza D. Current developments in reversible solid oxide fuel cells. Renewable and Sustainable Energy Reviews. 2016;**61**:155-174

[16] Zhang X, Chan SH, Li G, Ho HK, Li J, Feng Z. A review of integration strategies for solid oxide fuel cells. Journal of Power Sources. 2010;**195**(3): 685-702

[17] Cinti G, Baldinelli A, Di Michele A, Desideri U. Integration of solid oxide electrolyzer and Fischer-Tropsch: A sustainable pathway for synthetic fuel. Applied Energy. 2016;**162**:308-320

[18] Luo Y, Wu X, Shi Y, Ghoniem AF, Cai N. Exergy analysis of an integrated solid oxide electrolysis cell-methanation reactor for renewable energy storage. Applied Energy. 2018;**215**:371-383

[19] Kezibri N, Bouallou C. Conceptual design and modelling of an industrial scale power to gas-oxy-combustion power plant. International Journal of Hydrogen Energy. 2017;**42**(30):1-9

[20] Aziz M. Power generation from algae employing enhanced process integration technology. Chemical Engineering Research and Design. 2016; **109**:297-306

[21] Aziz M, Oda T, Kashiwagi T. Integration of energy-efficient drying in microalgae utilization based on enhanced process integration. Energy. 2014;**70**:307-316

[22] Darmawan A, Hardi F, Yoshikawa K, Aziz M, Tokimatsu K. Enhanced process integration of black liquor evaporation, gasification, and combined cycle. Applied Energy. 2017;**204**:1035-1042

[23] Udagawa J, Aguiar P, Brandon NP. Hydrogen production through steam electrolysis: Model-based steady state performance of a cathode-supported intermediate temperature solid oxide electrolysis cell. Journal of Power Sources. 2007;**166**(1):127-136

[24] Guo Q, Zhang J, Tian H. Recent advances in $CaSO_4$ oxygen carrier for chemical-looping combustion (Clc) process. Chemical Engineering Communications. 2012;**199**(11):1463-1491

[25] Fan L. Chemical Looping System for Fossil Energy Conversions. John Wiley & Sons; 2010. pp. 143-214

[26] Song Q et al. Chemical-looping combustion of methane with CaSO4 oxygen carrier in a fixed bed reactor. Energy Conversion and Management. 2008;**49**(11):3178-3187

Chapter 3

Exergy in Photovoltaic/Thermal Nanofluid-Based Collector Systems

Amin Farzanehnia and Mohammad Sardarabadi

Abstract

This chapter focuses on the exergy analysis of photovoltaic/thermal (PVT) systems using nanofluid. The PVT hybrid systems are designed to harness solar energy more efficiently. The thermodynamic theory of exergy in PVT systems is explained in details. The existing researches used various models to perform the exergy analysis for performance evaluation of the PVT systems. These models and formulations are compared with each other to achieve a widely used theory for a better comparison of the results. The exergy analysis is an effective tool to evaluate the performance of PVT systems. The exergy efficiency enhancement in PVT systems and the effect of nanofluid from the literature are presented. The literature survey suggests that the increase in the flow rate increases the exergy efficiencies in collector-based PVT. Using nanofluid as optical filters of solar radiation results in higher exergy efficiencies compared to collector-based PVT systems. According to the recent publications, the long-term thermophysical stability of nanofluid and cost-based exergy analysis still require further investigations.

Keywords: PVT, photovoltaic/thermal, solar energy, nanofluids, exergy, nanoparticles

1. Introduction

Solar energy is a proven alternative for fossil fuels due to its sustainability and availability. The ever-reducing investment costs of solar system installation made this technology highly popular. The use of solar energy is a great method for mitigating the environmental and health problems presented by nonrenewable energy sources in the long term. Photovoltaic (PV) systems and hybrid photovoltaic/thermal (PVT) systems are the major tools for solar radiation to electrical and thermal energy conversion. The PVT systems are a combination of PV and solar collector to produce electricity and heat simultaneously. These hybrid systems have the benefits of increasing the electrical efficiency of the PV, obtaining thermal energy, and avoiding thermal degradation of PV solar panels [1]. **Figure 1** shows the schematic of a PVT system. The diagram of the PVT system with a water-based sheet and tube thermal collector is presented in **Figure 2**. It is shown that by reducing 1°C in PV cell temperature, the electrical efficiency increases by 0.4–0.5% [2]. For amorphous silicon (a-Si), this increase is smaller about 0.25%/°C [3]. Additionally, the combination of solar collector and PV technologies has the

Figure 1.
The schematics of a PVT system.

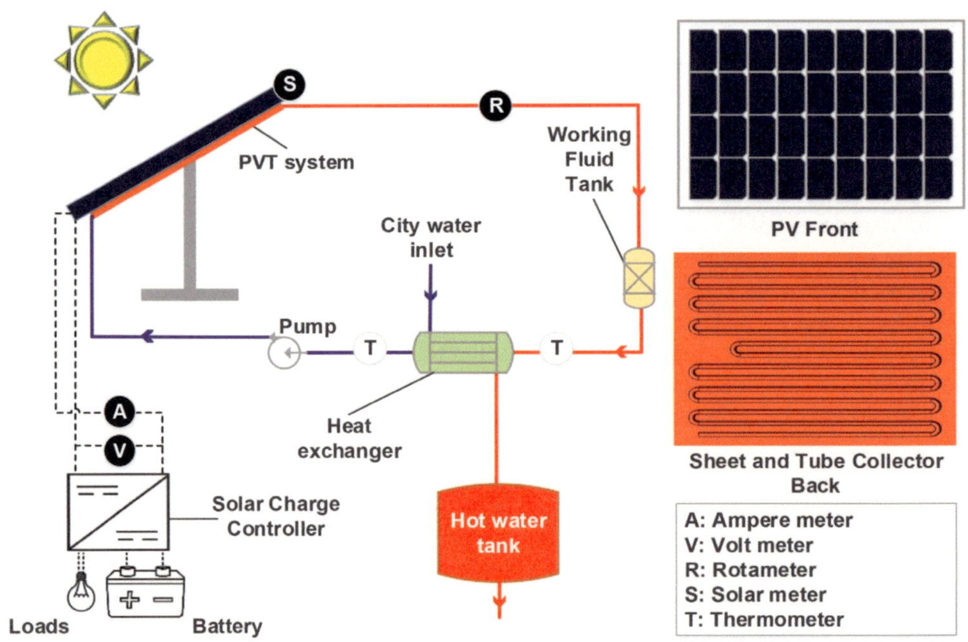

Figure 2.
The diagram of a PVT system with a water-based sheet and tube thermal collector.

advantages of reducing material usage, time of installation, and space required [4]. It is therefore necessary to evaluate the efficiency of PV and PVT systems to improve the design and usage of the systems, moreover, to help researches to make decisions on the usage and improvement of efficient systems.

Regarding the working fluid, PVT systems are classified under three groups of air-based PVT collector [5, 6], water-based PVT collector [7, 8], and a combination of air/water PVT collector [9, 10]. The PVT systems are mainly employed for low-temperature applications including space heating and air/water preheating in domestic buildings [11]. The type of working fluid is of primary significance to achieve energy-efficient systems compared to conventional fluids. The use of nanofluids due to their advanced thermal properties is gaining more and more attention [12]. Nanofluids are able to increase the efficiency of solar systems due to their augmented heat transfer properties such as thermal conductivity [13].

Exergy is an important means to evaluate the efficient use of the PV and PVT systems. The term exergy is defined as the maximum theoretical work potential of a system that interacts with an environment with constant conditions. In other words, exergy is defined as the energy that is available to be used. Therefore, one can evaluate the performance of a system by using the exergetic (second law) of thermodynamics.

2. Theory

The performance of PVT systems could be analyzed by the second law of thermodynamics (exergy analysis). In contrast to energy analysis, the exergy analysis takes the quality of energies into consideration. As mentioned before, the output energy of a PVT system is distributed over two forms of thermal and electrical energies. However, the quality of electrical energy is different from that of thermal energy. The electrical energy is equivalent to the available work, while only a part of thermal energy could be exploited as available work.

2.1 Energy analysis

Prior to performing the exergy analysis, the absorbed solar irradiation and output thermal and electrical power of PVT are required to be found. The PVT system is considered as control volume and is assumed to be in semi-steady state condition. The solar irradiation absorbed by PVT is calculated by [14]:

$$\dot{E}_{sun} = \dot{G_T}\, A_{PV}\, \tau_g\, \alpha_{cell} \tag{1}$$

where $\dot{G_T}$ is the total solar irradiation, A_{PV} is the PV area, τ_g is transmissivity of the glass layer over the PV module, and α_{cell} absorptivity of PV cells. The output energy from the thermal collector is given by:

$$\dot{E}_{th} = \dot{m}\, C_p (T_{out} - T_{in}) \tag{2}$$

In Eq. (2) the term $\dot{m}$ refers to the mass flow rate of the coolant fluid in the thermal collector of PVT system, C_p is the specific heat capacity, and the T_{in} and T_{out} are the inlet temperature and the outlet temperature of the coolant fluid, respectively.

In the condition where the coolant fluid is a nanofluid, the specific heat capacity of nanofluid is expressed as either Eq. (3) or (4) [15]:

$$C_{p.nf} = \varphi C_{p,np} + (1-\varphi) C_{p,np} \tag{3}$$

$$C_{p.nf} = \frac{\varphi (\rho C_p)_{np} + (1-\varphi)(\rho C_p)_{bf}}{\rho_{nf}} \tag{4}$$

where the subscripts nf, np, and bf denote nanofluid, nanoparticles, and base fluid, respectively. In the above equations, ρ is the density of the corresponding materials. The term φ is the volume fraction of nanoparticles which is given by:

$$\varphi = \left(\frac{m_{np}}{\rho_{np}}\right) \bigg/ \left(\frac{m_{np}}{\rho_{np}} + \frac{m_{bf}}{\rho_{bf}}\right) \tag{5}$$

where m_{np} and m_{bf} refer to the mass of nanoparticles and base fluid, respectively. The density of nanofluid ρ_{nf} in Eq. (4) is simply given by the two-phase mixture principle:

$$\rho_{nf} = \varphi\rho_{np} + (1-\varphi)\rho_{bf} \tag{6}$$

Equation (4) is proposed more in the literature than Eq. (3) due to the better agreement with the experimental results [15–17].

The main output of a PV module is the electrical energy output which is given by [18]:

$$\dot{E}_{el} = V_{oc} \times I_{sc} \times FF \tag{7}$$

where V_{oc}, I_{sc}, and FF are the open circuit voltage, short circuit current, and fill factor, respectively.

2.2 Exergy analysis

For better understanding, the diagram of exergy flows belonging to a PVT system is shown in **Figure 3**. The first step to investigate the performance of PVT system from exergy viewpoint is to consider the PVT system as a control volume. It is assumed that the system is in semi-steady condition.

The exergy balance of a PVT system is expressed as:

$$\sum\dot{Ex}_{in} = \sum\dot{Ex}_{out} + \sum\dot{Ex}_{dest} \tag{8}$$

where $\sum\dot{Ex}_{in}$ is inlet exergy, $\sum\dot{Ex}_{out}$ is the outlet exergy, and $\sum\dot{Ex}_{dest}$ is the exergy loss or destruction due to the irreversibility.

The term $\sum\dot{Ex}_{in}$ is the net input exergy rate. In solar systems such as PVT, the input energy is the solar radiation that reaches the system; therefore, the input exergy is equal to the exergy of incident solar irradiation to the system ($\dot{Ex}_{sun}$):

$$\sum\dot{Ex}_{in} = \dot{Ex}_{sun} \tag{9}$$

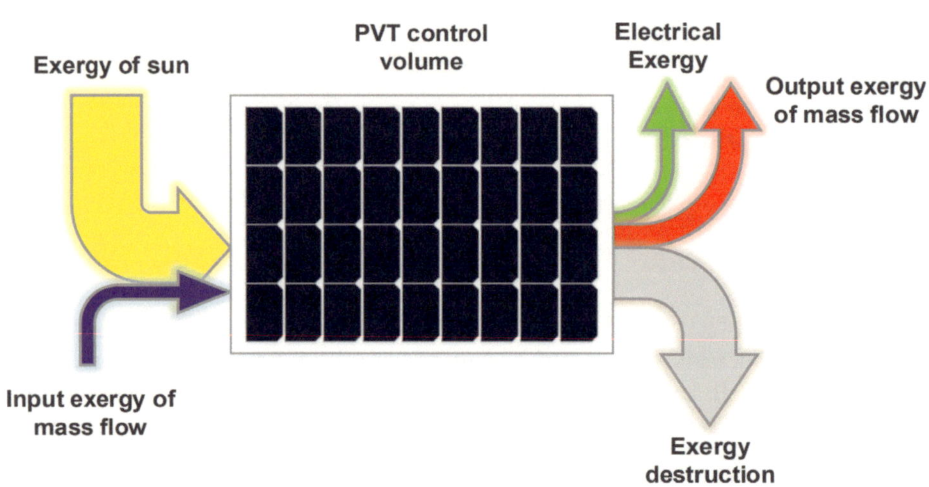

Figure 3.
The exergy flow diagram of a PVT system.

The net output exergy ($\sum \dot{Ex}_{out}$) of PVT systems consists of thermal exergy ($\dot{Ex}_{th}$) and electrical exergy ($\dot{Ex}_{el}$):

$$\sum \dot{Ex}_{out} = \dot{Ex}_{th} + \dot{Ex}_{el} \tag{10}$$

Considering that the thermal exergy is equal to the difference of flow exergy at the outlet and inlet of the collector ($\dot{Ex}_{th} = \dot{Ex}_{mass.out} - \dot{Ex}_{mass.in}$), the exergy balance becomes (Eqs. 8–10):

$$\dot{Ex}_{sun} = \dot{Ex}_{th} + \dot{Ex}_{el} + \dot{Ex}_{dest} \tag{11}$$

Many methods have been proposed to evaluate the exergy of the solar irradiation. The three following equations are the most commonly used equations for the exergy of absorbed solar irradiation by the PVT proposed, respectively, by Jeter [19], Spanner [20], and Petala [21]:

$$\dot{Ex}_{sun} = \dot{E}_{sun}\left(1 - \frac{T_{amb}}{T_{sun}}\right) \tag{12}$$

$$\dot{Ex}_{sun} = \dot{E}_{sun}\left(1 - \frac{4T_{amb}}{3T_{sun}}\right) \tag{13}$$

$$\dot{Ex}_{sun} = \dot{E}_{sun}\left[1 - \frac{4T_{amb}}{3T_{sun}} + \frac{1}{3}\left(\frac{T_{amb}}{T_{sun}}\right)^4\right] \tag{14}$$

where T_{amb} is the ambient temperature and T_{sun} is the surface temperature of the sun as a blackbody. Although it is hotter inside, the sun temperature could be estimated at its surface where the emissions occur and is approximated as a blackbody at 5770 K. The results from Eqs. (12)–(14) differ from each other less than 2% [22]. However, the literature review by Kalogirou [23] indicates that Eq. (14) (Petala equation) is proposed and used more often than the other two equations.

The output thermal exergy of the PVT system ($\dot{Ex}_{th}$) is given by both Eqs. (15) and (16) [24, 25]:

$$\dot{Ex}_{th} = \dot{E}_{th}\left(1 - \frac{T_{amb}}{T_{out}}\right) \tag{15}$$

$$\dot{Ex}_{th} = \dot{m}\left[(h_{out} - h_{in}) - T_{amb}(s_{out} - s_{in})\right] = \dot{m}\ C_p\left[(T_{out} - T_{in}) - T_{amb}\ln\left(\frac{T_{out}}{T_{in}}\right)\right] \tag{16}$$

where h_{out} and s_{out} are the fluid enthalpy and entropy at the outlet of the collector. Similarly, h_{in} and s_{in} are the fluid enthalpy and entropy at the inlet of the collector.

Equation (15) derives the exergy based on a series of imaginary heat engine that operates between source temperature of T_{out} and the sink temperature of T_{amb}. Thus, the thermal exergy is the available work extracted from a Carnot efficiency heat engine between the outlet fluid temperature and ambient temperature. Equation 16 is often encountered in the literature [24, 26].

The electrical energy output is equivalent to the electrical exergy. The electrical energy could be 100% converted to work. Therefore, the electrical exergy of a passive PVT is expressed as:

$$\dot{Ex}_{el} = \dot{E}_{el} \tag{17}$$

However, the electrical exergy of an active PVT system is defined as the difference between electrical power and the required pumping power Eq. (18) [27]. Nevertheless, some studies do not consider the pumping power and simply use Eq. (17) [11]:

$$\dot{Ex}_{el} = \dot{E}_{el} - \dot{E}_{pump} \tag{18}$$

where $\dot{E}_{pump}$ is the electrical power consumption of the pump, which can be expressed as [28]:

$$\dot{E}_{pump} = \frac{\dot{m}\Delta P}{\rho \eta_p} \tag{19}$$

where η_p is the pump efficiency. Thermal and electrical exergy efficiencies based on second law are given as:

$$\varepsilon_{th} = \frac{\dot{Ex}_{th}}{\dot{Ex}_{sun}} \times 100 \tag{20}$$

$$\varepsilon_{el} = \frac{\dot{Ex}_{el}}{\dot{Ex}_{sun}} \times 100 \tag{21}$$

The overall exergy efficiency could be given as follows:

$$\varepsilon_{total} \cong \frac{\dot{Ex}_{th} + \dot{Ex}_{el}}{\dot{Ex}_{sun}} = \frac{\int_{t_2}^{t_1} \left(A_c \dot{Ex}''_{th} + A_{PV} \dot{Ex}''_{el}\right) dt}{A_c \int_{t_2}^{t_1} \left(\dot{Ex}''_{sun}\right) dt} = \varepsilon_{th} + r\varepsilon_{el} \tag{22}$$

where A_{PV} and A_c are the PV panel and collector areas, respectively. $\dot{Ex}''_{th}$ is the rate of the output thermal exergy per unit area of collector, $\dot{Ex}''_{el}$ is the rate of electrical exergy per unit area of PV module, and $\dot{Ex}''_{sun}$ is the rate of solar irradiation exergy per unit area of collector. The term r is the packing factor and defined as the area of PV panel to the collector ($r = A_{PV}/A_c$). Therefore, when the packing factor is equal to 1, the overall exergy is simply the sum of thermal and electrical exergy efficiencies.

It worth mentioning that the exergy destruction term $\dot{Ex}_{dest}$ in Eqs. (8) and (11) is due to heat transfer losses and frictional losses in the collector tube. In the active PVT systems, in addition to heat transfer exergy loss, there is another exergy loss due to head or frictional loss in the collector tube, which can be derived by:

$$\dot{E}_{loss,fr} = \frac{\dot{m}\Delta P}{\rho} \frac{T_{amb} . \ln\left({T_{out}}/{T_{in}}\right)}{\left(T_{out} - T_{in}\right)} \tag{23}$$

where ΔP is the pressure drop along the collector and ρ is the fluid density. The rate of entropy generation by irreversibility in the control volume could be calculated as:

$$\dot{S}_{gen} = \frac{\dot{Ex}_{dest}}{T_{amb}} \tag{24}$$

where the $\dot{Ex}_{dest}$ is the total rate of exergy destruction which is the sum of exergy destruction due to the heat transfer loss and due to pressure drop in the collector tube and is given by Eq. (11).

3. Development of nanofluid-based PVT

The PVT system is employed to produce electrical and thermal energies simultaneously. The study of PVT water-based collectors has been limited in the literature due to the small market size. Also, most literature is focused on custom-made PVT systems [4]. However, now there are various configurations of commercialized PVT available, and these systems became popular [29, 30]. As the market size increases, it is necessary to evaluate and enhance the performance of PVT systems. Recently, the nanofluids are used in the PVT systems as working fluid or as optical filters resulting in increased efficiency of the systems. There are several reviews published on the topic of nanofluid PVT in the last couple of years such as by Said et al. [31], by Yazdanifard et al. [27], by Al-Shamani et al. [12], and by Ali et al. [32].

Sardarabadi et al. [14] performed an experimental investigation on silica/water nanofluid PVT systems based on first and second laws of thermodynamics. The mass flow rates 20, 30, and 40 L/h were studied, and the optimum mass flow rate for the working fluid was determined. The thermal exergy efficiency of the system was much smaller than the electrical efficiency. This was attributed to the small temperature difference between outlet and ambient temperatures. The results indicated that using thermal collector with the PV module increases the overall efficiency. Also using nanofluid enhances the energy efficiency of the system. However, comparing the results from the second law to the first law of thermodynamics, it was shown that although the thermal and overall efficiency from the first law viewpoint is high, the thermal exergetic and therefore overall efficiency were low (**Table 1**). This was due to the low-quality (low temperature) thermal energy in PVT systems.

Moradgholi et al. [35] studied two-phase closed thermosyphon (TPCT) PVT system using Al_2O_3/methanol nanofluid as the working fluid. They studied the effects of various mass fractions of nanoparticles 1, 1.5, and 2 wt%, and also the effect of filling ratio (the working fluid volume to the evaporating section volume) was studied. The optimum values of thermal and electrical performance of the system were obtained at the mass concentration of 1.5 wt% and the filling ratio of 50%. The average overall exergy efficiency was 11.7, 12.5, and 12.7% for PV module, PVT module with base fluid, and PVT module with nanofluid, respectively.

Sardarabadi et al. [34] studied the effects of both ZnO/water nanofluid and phase-change material (PCM) as a coolant in photovoltaic thermal systems. They used a PV module as a reference point and performed energy and exergy analysis on PV, PVT, and PVT with PCM systems with water and ZnO/water nanofluid as working fluids. An increase of nearly twice in the thermal exergy output was observed using PCM in PVT modules. This was also shown in other studies [26] and is because the heat generated in PV cells is absorbed in PCM and could be used as thermal energy. Also, the overall exergy efficiency of 10% was found for the PV module, whereas the PVT module and PVT module with PCM using nanofluid had an efficiency of 12.29 and 13.42%, respectively.

Sardarabadi et al. [24] studied the effects of using metal-oxides/water nanofluids on a PVT system from energy and exergy viewpoints. The Al_2O_3, TiO_2, and ZnO nanoparticles with the mass fraction of 0.2 wt% were considered. It was shown that the ZnO/water nanofluid had the highest energy and exergy efficiencies and TiO_2/water had the highest electrical exergy efficiency than other systems. The average overall exergy efficiency for the PV, PVT/water, PVT/TiO_2, PVT/ZnO, and PVT/Al_2O_3 was 10.29, 11.56, 11.93, 12.17, and 11.88%, respectively.

Brekke et al. [36] proposed a performance model of a concentrating hybrid PVT system utilizing selective spectral nanofluid absorption. The proposed system used nanofluid to absorb the portion of the solar spectrum not efficiently exploited by

Ref.	Working fluid type	Concentrations	Flow rate	Amb. temp. (°C)	Exergy efficiencies (%)		
					Thermal	Electrical	Overall
[14]	Without cooling	—	—	33	0	11.53	11.53
	Pure water	0	30 L/h		1.23	12.47	13.54
	SiO_2/water	1 wt%			1.48	12.57	13.85
	SiO_2/water	3 wt%			1.68	12.59	14.02
[33]	Ag/water	2 and 4 wt%	0.0085, 0.016, and 0.029 kg/s corresponding to laminar, transient, and turbulent regimes	25			4 wt%, turbulent regime: 50 and 30% improvements in exergy efficiency compared to water coolant
[34]	Without cooling	—	—	—	0	10.9	10.9
	PVT pure water	0	30 kg/h	—	0.50	11.73	12.23
	PVT ZnO/ water	0.2 wt%			0.51	11.78	12.29
	PVT/PCM Pure water	0			0.87	12.30	13.17
	PVT/PCM ZnO/water	0.2 wt%			1.08	12.35	13.42
[35]	Without cooling	—	—	—	—	—	11.7
	Pure water	—	15 (l/min)		—	—	12.5
	Al_2O_3/ methanol	1, 1.5, and 2 wt %			—	—	12.7 for the optimum conditions
[24]	Without cooling			—	0	10.29	10.29
	Pure water		30 kg/h		0.72	10.84	11.56
	TiO_2/water	0.2 wt%			0.91	11.02	11.93
	ZnO/water	0.2 wt%			1.18	10.99	12.17
	Al_2O_3/ water	0.2 wt%			1.01	10.87	11.88
[36]	Au/ Duratherm S	—	0.05 kg/s	19	—	—	41.3
	ITO/ Duratherm S				—	—	42.3
[37]	Ag/water	0.001–1.5 vol.%	0.08 kg/s	—	—	—	By increasing nanofluid volume fraction for GaAs cells: 24.2–30% For Si cells: 19.8–24.4%

Table 1.
List of studies on the PVT nanofluid collector-based systems.

the PV module. Two common PV cell materials of crystalline silicon (c-Si) and gallium arsenide (GaAs) were used in their numerical study. The heat transfer fluid of Duratherm S and gold nanoparticles were considered for c-Si, and indium tin oxide (ITO) nanoparticles were considered for GaAs PV module. The GaAs exhibited higher exergy efficiency due to higher PV efficiency of this cell; however, this results in the reduction of thermal exergy percentage of these cells because a smaller portion of thermal energy is absorbed by the spectral fluid. The results of exergy efficiency of previous studies are summarized in **Table 1**.

4. Conclusion

This chapter addresses the exergy in photovoltaic/thermal systems that contain nanofluid-based collector. These systems provide both thermal and electrical energies. The comprehensive theory of exergy analysis in these systems is elaborated. It is shown that existing researches used various models to perform the exergy analysis for performance evaluation of the photovoltaic/thermal systems. These models are compared with each other to achieve a widely used theory for a better comparison of the results. The literature survey on nanofluid in PVT indicates that the overall exergy efficiency is generally in the range of 10–14% for PVT collectors. The increasing flow rate and transition to turbulent flow increase exergy efficiency. When using nanofluid as optical filters, higher exergy efficiencies were observed. The performance of PVT systems could be analyzed by exergy analysis. Despite the benefits of nanofluid PVT system, barriers to the development of these systems are agglomeration, required pumping power, and pipe erosions. Additionally, the use of nanofluid collectors and optical filters brings much cost to the system. Therefore future investigation is required on exergy-based cost analysis, nanofluid stability, and optimization of PVT systems.

Nomenclature

A	area (m^2)
C_P	specific heat capacity ($J\ kg^{-1}\ K^{-1}$)
E	energy (J)
Ex	exergy (J)
$\dot{E}$	power (W)
FF	fill factor
$\dot{G}$	solar irradiation rate ($W\ m^{-2}$)
I	electrical current (A)
$\dot{m}$	mass flow rate ($kg\ s^{-1}$)
k	thermal conductivity ($W\ m^{-1}\ K^{-1}$)
P	pressure (Pa)
T	temperature (K)
V	velocity (m/s)
r	packing factor

Greeks

α	absorptivity
ε	exergy efficiency (%)

ρ	density (kg m^{-3})
τ	transmissivity
ϕ	nanoparticles volume fraction

Subscripts

amb	ambient
bf	base-fluid
dest	destruction
el	electrical
in	inlet
n	nanoparticle
nf	nanofluid
oc	open circuit
out	outlet
ov	overall
t	total
sc	short circuit
th	thermal

Author details

Amin Farzanehnia[1] and Mohammad Sardarabadi[2]*

1 Department of Mechanical Engineering, Ferdowsi University of Mashhad, Mashhad, Iran

2 Department of Energy, Quchan University of Technology, Quchan, Iran

*Address all correspondence to: m.sardarabadi@yahoo.com

References

[1] Al-Musawi AIA, Taheri A, Farzanehnia A, Sardarabadi M, Passandideh-Fard M. Numerical study of the effects of nanofluids and phase-change materials in photovoltaic thermal (PVT) systems. Journal of Thermal Analysis and Calorimetry. 2018

[2] Al-Waeli AH, Sopian K, Chaichan MT, Kazem HA, Hasan HA, Al-Shamani AN. An experimental investigation of SiC nanofluid as a base-fluid for a photovoltaic thermal PV/T system. Energy Conversion and Management. 2017;**142**:547-558

[3] Nižetić S, Giama E, Papadopoulos A. Comprehensive analysis and general economic-environmental evaluation of cooling techniques for photovoltaic panels, Part II: Active cooling techniques. Energy Conversion and Management. 2018;**155**:301-323

[4] Good C. Environmental impact assessments of hybrid photovoltaic–thermal (PV/T) systems–A review. Renewable and Sustainable Energy Reviews. 2016;**55**:234-239

[5] Sarhaddi F, Farahat S, Ajam H, Behzadmehr A, Mahdavi Adeli M. An improved thermal and electrical model for a solar photovoltaic thermal (PV/T) air collector. Applied Energy. 2010;**87**:2328-2339

[6] Su D, Jia Y, Alva G, Liu L, Fang G. Comparative analyses on dynamic performances of photovoltaic–thermal solar collectors integrated with phase change materials. Energy Conversion and Management. 2017;**131**:79-89

[7] Yazdanpanahi J, Sarhaddi F, Mahdavi Adeli M. Experimental investigation of exergy efficiency of a solar photovoltaic thermal (PVT) water collector based on exergy losses. Solar Energy. 2015;**118**:197-208

[8] Fudholi A, Sopian K, Yazdi MH, Ruslan MH, Ibrahim A, Kazem HA. Performance analysis of photovoltaic thermal (PVT) water collectors. Energy Conversion and Management. 2014;**78**:641-651

[9] Yu B, Jiang Q, He W, Liu S, Zhou F, Ji J, et al. Performance study on a novel hybrid solar gradient utilization system for combined photocatalytic oxidation technology and photovoltaic/thermal technology. Applied Energy. 2018;**215**:699-716

[10] Othman MY, Hamid SA, Tabook MAS, Sopian K, Roslan MH, Ibarahim Z. Performance analysis of PV/T Combi with water and air heating system: An experimental study. Renewable Energy. 2016;**86**:716-722

[11] Bayrak F, Abu-Hamdeh N, Alnefaie KA, Öztop HF. A review on exergy analysis of solar electricity production. Renewable and Sustainable Energy Reviews. 2017;**74**:755-770

[12] Al-Shamani AN, Yazdi MH, Alghoul M, Abed AM, Ruslan MH, Mat S, et al. Nanofluids for improved efficiency in cooling solar collectors–A review. Renewable and Sustainable Energy Reviews. 2014;**38**:348-367

[13] Al-Shamani AN, Sopian K, Mat S, Hasan HA, Abed AM, Ruslan M. Experimental studies of rectangular tube absorber photovoltaic thermal collector with various types of nanofluids under the tropical climate conditions. Energy Conversion and Management. 2016;**124**:528-542

[14] Sardarabadi M, Passandideh-Fard M, Heris SZ. Experimental investigation of the effects of silica/water nanofluid on PV/T (photovoltaic thermal units). Energy. 2014;**66**:264-272

[15] Mahian O, Kianifar A, Kleinstreuer C, Al-Nimr MA, Pop I, Sahin AZ, et al. A review of entropy generation in nanofluid flow. International Journal of Heat and Mass Transfer. 2013;**65**: 514-532

[16] Mahian O, Kianifar A, Sahin AZ, Wongwises S. Entropy generation during Al_2O_3/water nanofluid flow in a solar collector: Effects of tube roughness, nanoparticle size, and different thermophysical models. International Journal of Heat and Mass Transfer. 2014;**78**:64-75

[17] Khanafer K, Vafai K. A critical synthesis of thermophysical characteristics of nanofluids. International Journal of Heat and Mass Transfer. 2011;**54**:4410-4428

[18] Hosseinzadeh M, Salari A, Sardarabadi M, Passandideh-Fard M. Optimization and parametric analysis of a nanofluid based photovoltaic thermal system: 3D numerical model with experimental validation. Energy Conversion and Management. 2018;**160**: 93-108

[19] Jeter SM. Maximum conversion efficiency for the utilization of direct solar radiation. Solar Energy. 1981;**26**: 231-236

[20] Spanner DC. Introduction to Thermodynamics. London and New York: Academic Press; 1964

[21] Petela R. Exergy of heat radiation. Journal of Heat Transfer. 1964;**86**: 187-192

[22] Chow TT, Pei G, Fong K, Lin Z, Chan A, Ji J. Energy and exergy analysis of photovoltaic–thermal collector with and without glass cover. Applied Energy. 2009;**86**:310-316

[23] Kalogirou SA, Karellas S, Braimakis K, Stanciu C, Badescu V. Exergy analysis of solar thermal collectors and processes. Progress in Energy and Combustion Science. 2016;**56**:106-137

[24] Sardarabadi M, Hosseinzadeh M, Kazemian A, Passandideh-Fard M. Experimental investigation of the effects of using metal-oxides/water nanofluids on a photovoltaic thermal system (PVT) from energy and exergy viewpoints. Energy. 2017;**138**:682-695

[25] Ghadiri M, Sardarabadi M, Pasandideh-fard M, Moghadam AJ. Experimental investigation of a PVT system performance using nano ferrofluids. Energy Conversion and Management. 2015;**103**:468-476

[26] Hosseinzadeh M, Sardarabadi M, Passandideh-Fard M. Energy and exergy analysis of nanofluid based photovoltaic thermal system integrated with phase change material. Energy. 2018;**147**: 636-647

[27] Yazdanifard F, Ameri M, Ebrahimnia-Bajestan E. Performance of nanofluid-based photovoltaic/thermal systems: A review. Renewable and Sustainable Energy Reviews. 2017;**76**: 323-352

[28] Zainine MA, Mezni T, Dakhlaoui MA, Guizani A. Energetic performance and economic analysis of a solar water heating system for different flow rates values: A case study. Solar Energy. 2017; **147**:164-180

[29] Axaopoulos PJ, Fylladitakis ED. Performance and economic evaluation of a hybrid photovoltaic/thermal solar system for residential applications. Energy and Buildings. 2013;**65**:488-496

[30] Lamnatou C, Chemisana D. Photovoltaic/thermal (PVT) systems: A review with emphasis on environmental issues. Renewable Energy. 2017;**105**: 270-287

[31] Said Z, Arora S, Bellos E. A review on performance and environmental

effects of conventional and nanofluid-based thermal photovoltaics. Renewable and Sustainable Energy Reviews. 2018; **94**:302-316

[32] Ali HM, Shah TR, Babar H, Khan ZA. Microfluidics and Nanofluidics. IntechOpen; 2018

[33] Aberoumand S, Ghamari S, Shabani B. Energy and exergy analysis of a photovoltaic thermal (PV/T) system using nanofluids: An experimental study. Solar Energy. 2018;**165**:167-177

[34] Sardarabadi M, Passandideh-Fard M, Maghrebi MJ, Ghazikhani M. Experimental study of using both ZnO/water nanofluid and phase change material (PCM) in photovoltaic thermal systems. Solar Energy Materials and Solar Cells. 2017;**161**:62-69

[35] Moradgholi M, Nowee SM, Farzaneh A. Experimental study of using Al_2O_3/methanol nanofluid in a two phase closed thermosyphon (TPCT) array as a novel photovoltaic/thermal system. Solar Energy. 2018;**164**:243-250

[36] Brekke N, Dale J, DeJarnette D, Hari P, Orosz M, Roberts K, et al. Detailed performance model of a hybrid photovoltaic/thermal system utilizing selective spectral nanofluid absorption. Renewable Energy. 2018;**123**:683-693

[37] Hassani S, Taylor RA, Mekhilef S, Saidur R. A cascade nanofluid-based PV/T system with optimized optical and thermal properties. Energy. 2016;**112**:963-975

Section 3

Energy Management System

Chapter 4

Electrical Vehicle-Assisted Demand Side Energy Management

Xing Luo, Xu Zhu and Eng Gee Lim

Abstract

The recent development of electrical vehicles (EVs) offers vast benefits not only in environmental protection and economics but also in demand response (DR). Employing EVs in load scheduling enables householders to help alleviate the network load burden while reducing their own electric bills. In this chapter, innovative EV-assisted DR strategies with an EV auxiliary power supply (APS) model and a neighbor energy sharing (NES) model are proposed, to jointly optimize the load distribution for both individual household and multi-household network via vehicle-to-home (V2H) and vehicle-to-neighbor (V2N)connections, respectively. The proposed DR strategies take account of the comprehensive impacts of EV charging behaviors, user preferences, distributed generation, and load priority. The effectiveness of the proposed energy management solutions is verified by numerical results in terms of load balancing and cost reduction. The proposed DR strategies also significantly outperform the previous approaches.

Keywords: electrical vehicle, demand response, energy management, auxiliary power supply, energy sharing

1. Introduction

Among a variety of innovative technologies in the twenty-first century, demand response (DR) has been regarded as a promising long-term solution to improving energy efficiency and reducing energy wastage. It also plays a significant role in both balancing energy supply and demand and enhancing the reliability in smart grid [1–3]. The basic concept of DR management is to reduce or shift the demand for electricity during peak periods in response to dynamic pricing (DP) or other forms of financial incentive, thus achieving the aim of saving electric bills for customers. In other ways, it is also beneficial for power grid as it offers an effective solution to average the power usage in certain periods to alleviate the load burden of the power grid [1–4].

Meanwhile, electric vehicles (EVs) are becoming a trend in the next generation of transportation due to their economic and environmental benefits and the rapid advance of rechargeable battery technology [5–7]. Along with the worldwide application of DP, an increasing adoption of EVs in residences brings about both opportunities and challenges for smart grid. Residences with EVs consume more electricity and react more elastically to electricity price [8]. According to the report provided by the US Energy Information Association [9], the fast charging of an EV is equivalent to about 120 houses coming on line for half an hour, which is a severe issue to the power grid. On the other hand, the usage of EVs as energy storage units via vehicle to home (V2H) offers an effective solution to load shaping at demand side. In addition to this, the surplus energy of EVs can be delivered to neighbor via vehicle to neighbor (V2N) if it is enabled. Hence, householders are able to participate in load scheduling and may have multiple options in energy allocation.

The importance of DR cooperating with EVs increases, since EVs become prevalent recently. Considering the flexible energy storage purpose of EVs, more up-to-date DR strategies that take the behaviors of EVs into account are required. The implementation of DR with EVs requires efficient energy distribution management and high-performance batteries as basis. Moreover, DP provides a basic control signal to optimally schedule the charging and discharging of EVs, by minimizing the overall cost.

Compared with the conventional energy storage system (ESS) and other energy production facilities, the utilization of EV as a temporary power source has advantages in employing flexibility and economic efficiency [10]. It does not expect extra investment besides the daily used EVs. Meanwhile, the power sharing is enabled from the V2N connection. The surplus energy of EVs can be shared to neighbors during peak price time and benefit for both sides. Therefore, the DR strategy with EVs holds wide prospects in practice not only for an individual household but also for the multi-household network.

Much research has been conducted on demand response, and there are many popular DR strategies considering EV impacts being presented in literature. For example, in [11], an optimization framework-based DR program was proposed, with high penetration of EVs and storage systems from residential customer's perspective as well as utility company's perspective. The simulation results showed that the appropriate scheduling has benefits for both customers and suppliers. In [12], authors focused on EVs' charging behaviors based on the collected data from EV charging session, and different types of charging behaviors were derived. Nonetheless, the specific DR program with the proposed charging profiles has not been declared. To analyze the potential usage of EVs in power grid, the optimal time of EVs' charging and discharging was explored in [13]. However, all the mentioned studies above are limited to the operation of a single user and fail to attempt the scheduling of EVs among a group of households in DR program.

Moreover, authors in [14] proposed an algorithm for EVs' scheduling in DR to optimize the peak demand. The optimization problem is studied in a game framework. However, other electric appliances have not been considered in this work. In [15], an intelligent preemptive DR management using a building energy management system was proposed to better schedule the energy consumption within buildings. In this work, dynamic EV charging scheduling, priority-based load shedding, and air-conditioning system were accounted. Authors in [16] presented an optimal behavior of plug-in EV parking lots in the energy and reserve market. Both price-based and incentive-based DR programs were developed, and uncertainties of plug-in EVs were also considered by using the stochastic programming approach. In addition to these, a number of interesting DR programs coordinating with EVs are also described in [17–19].

In this chapter, we propose two innovative EV-assisted DR strategies with an EV auxiliary power supply (EV-APS) model and a neighbor energy sharing (NES) model, to jointly optimize the load distribution for both individual household and multi-household network via vehicle-to-home (V2H) and vehicle-to-neighbor (V2N) connections, respectively. Compared with the previous research, the main contributions of this work are:

1. Two significant EV-assisted DR strategies for domestic appliance scheduling are designed and implemented to different scales of households (individual

household network and multi-household network) in order to alleviate the load burden for the grid and save electric bills for householders, simultaneously.

2. For the individual household network, EV is utilized as an auxiliary power supply (APS) for energy consumption of home appliances on special occasions. An EV-APS model-based DR strategy is proposed.

3. For the multi-household network, an EV-assisted DR strategy including a neighbor energy sharing (NES) model for a residential network with different types of EVs installed at consumers' premise is developed. The surplus EVs' energy distribution is enabled via vehicle-to-home (V2H) and vehicle-to-neighbor (V2N) connections in this chapter. The NES-based DR strategy is valid and effective not only for an independent household but also for a multi-household residential network, which can satisfy broader requirements compared with conventional DR strategies in literature. The energy trading policy in neighborhood is also declared.

4. Comprehensive affecting factors (e.g., EV behaviors, user preferences, load scheduling priorities, etc.) are considered in scheduling for both EV-assisted DR strategies. The effectiveness of the proposed DR strategies is verified by numerical results, which demonstrate that our approaches significantly outperform the methods in literature in terms of load balancing and electricity cost reduction.

2. Electrical vehicle-assisted demand response strategy for individual household

An innovative electrical vehicle (EV)-assisted demand response strategy for load scheduling within an individual household is illustrated in this section. An EV auxiliary power supply (EV-APS) model is presented first. Afterwards, the system models are introduced mathematically. At last, the problem formulation and optimization method are proposed.

2.1 EV-APS demand response network

The schematic diagram of the proposed DR strategy with the EV-APS model is shown in **Figure 1**. Specifically, householders buy electricity from the power grid for the daily usage including EV charging under a dynamic pricing (DP) tariff. Normally, domestic appliances are directly powered by the main power grid. However, as an interim energy storage unit, EV is able to supply power for the household appliances (HAs) in auxiliaries on appropriate occasions, especially in high-price periods. The time of activating EV-APS is dependent on the instructions from the smart controller.

In addition, the smart controller plays the role as a supervisor in the system network. It regulates the energy sources supplying and the operating time of the household appliances based on real-time load information which is received from the smart meter and other signals (e.g., DP, EV status, load priority, etc.).

Moreover, more than 15 types of household appliances will be used generally in domestic homes every day. Considering the operating characteristic of each appliance, it is not necessary to schedule all of them via DR programs. Hence, in accordance with the device operating characteristics, the household appliances can be classified into different scenarios. In this chapter, household appliances are defined and sorted into two main scenarios as follows:

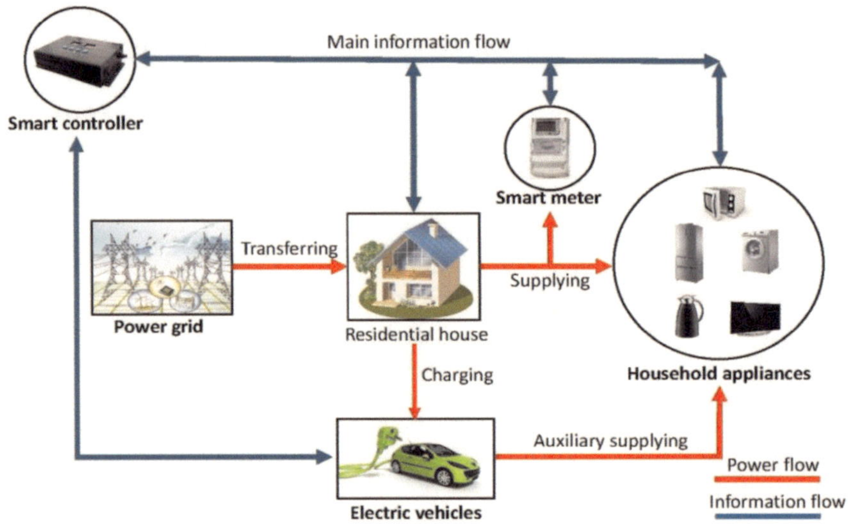

Figure 1.
Schematic diagram of an EV-APS model-based DR strategy for an individual household.

1. Critical scenario (CS): CS contains the appliances that have to be used at a specified time or cannot be scheduled. Examples include lightings, TV, laptop, etc.

2. Flexible scenario (FS): FS contains the appliances that can be powered on with a tolerable delay and have a flexible operating time. Hot water tank and washer are typical representatives in FS.

According to the sorting scheme above, 16 frequently used appliances including EVs are listed and classified with different types of jobs. They can be sorted as follows:

1. CS appliances: refrigerator, water dispenser, toaster, microwave oven, lights, electric cooker, electric kettle, TV, PC, hair drier, and cleaner

2. FS appliances: dish-washing machine, hot water tank, washer, drying machine, and EVs

2.2 System models

In this subsection, the formulation of the EV-APS DR strategy consisting of the main power supply model and auxiliary power supply model is illustrated mathematically.

2.2.1 Main power supply model

First, we define variables W_t^{grid} and P_t^{grid} as the total energy consumption and the total load power on grid at time t, respectively. Afterwards, the main power supply model with the corresponded constraints can be presented as

$$W_t^{\text{grid}} = \int_{T_{\text{in}}}^{T_{\text{term}}} P_t^{\text{grid}} \cdot d(t) \tag{1}$$

$$P_t^{\text{grid}} = P_t^{\text{HA}} + P_t^{\text{EV, c}} - P_t^{\text{EV, d}} \tag{2}$$

$$P_t^{\text{HA}} = \sum_{j=1}^{n} P_{t,j}^{\text{CS}} + \varepsilon_i \cdot \sum_{i=1}^{m} P_{t,i}^{\text{FS}} \tag{3}$$

Subject to

$$\forall t \in [T_{\text{in}}, T_{\text{term}}], P_t^{\text{grid}} \leq P_{\max}^{\text{grid}} \tag{4}$$

$$P_t^{\text{EV, c}} = 0, \text{if} P_t^{\text{EV, d}} > 0 \tag{5}$$

$$P_t^{\text{EV, d}} = 0, \text{if} P_t^{\text{EV, c}} > 0 \tag{6}$$

Equation (1) indicates that the total energy consumption (W_t^{grid}) is equal to the integral of total power (P_t^{grid}) through the time that is between initial time T_{in} and the terminate time T_{term}. Equation (2) illustrates the relationships between the total power and each power-consumed component. Variable P_t^{HA} denotes the load power consumed by the household appliances at time t. Variables $P_t^{\text{EV, c}}$ and $P_t^{\text{EV, d}}$ represent the power rates of the EV charging and discharging, respectively.

Additionally, as it is shown in Eq. (3), P_t^{HA} consists of the power cost by CS appliances ($P_{t,j}^{\text{CS}}$) and FS appliances ($P_{t,i}^{\text{FS}}$), where j and i represent the index of the appliances. The ε parameters have small positive values (e.g., $1{+}e^{-8}$, $1{+}2e^{-8}$, and $1{+}3e^{-8}$) that are determined by assumptions (the total power of appliances is not affected). This setting meets the requirement of having a priority according to user preferences in scheduling FS appliances. The smaller value of ε indicates a higher priority in the scheduling process by DR programs.

In spite of that, $P_{\max}^{\text{grid}}$ is proposed in Eq. (4) as a constraint to limit the maximum power rate on grid at time t for the safety and power distribution considerations. Further, constraints in Eqs. (5) and (6) express that the battery charging and discharging cannot be executed simultaneously; otherwise, the battery will be damaged to a certain extent.

2.2.2 Auxiliary power supply model

Determining the EV-APS model requires sufficient knowledge from previous research. According to the investigation of the current EV market, **Table 1** illustrates the core parameters of five major brands of EVs around the world.

Manufacturer and model	Battery capacity (kWh)	Discharging power (kW)	Driving range per charge (miles)
Tesla, Model S (EV)	60	3.0	273
BYD, Tang 100 (HEV)	23	3.3	63
BMW, i3 (EV/HEV)	33	2.5	114
GM, Chevrolet Bolt (EV)	60	—	283
Nissan, Leaf (EV)	30	—	107

Table 1.
Major brand of EVs in current market.

The parameters include the maximum battery capacity $W^{EV,\max}$, the discharging power $P_t^{EV,d}$, and the maximum driving range per full charge.

Moreover, multiple charging schemes are provided for each EV. **Table 2** shows the relevant charging schemes of Tesla Model S which will be used in simulations. It can be seen that the charging power $P_t^{EV,c}$ plays an important role in the grid due to the high power rate of battery charging.

Further, variables $W^{EV,(1)}$ and $W^{EV,(2)}$ are defined as the initial energy storage when people leave home in the morning of the first day and the second day, respectively. Therefore, the EV auxiliary power supply model can be proposed as follows:

$$W^{EV,rem} = W^{EV,(1)} - W^{EV,trip} \tag{7}$$

$$W^{EV,trip} = \frac{D^{trip}}{D^{max}} \cdot W^{EV,\max} \tag{8}$$

$$W^{EV,(2)} = W^{EV,rem} + W^{EV,c} - W^{EV,d} \tag{9}$$

$$W^{EV,c} = \eta_1 \cdot \int_{T_{c,b}}^{T_{c,e}} P_t^{EV,c} \cdot d(t) \tag{10}$$

$$W^{EV,d} = \eta_2 \cdot \int_{T_{d,b}}^{T_{d,e}} P_t^{EV,d} \cdot d(t) \tag{11}$$

Subject to

$$\forall t, W^{EV,\min} \leq W^{EV,rem} \leq W^{EV,\max} \tag{12}$$

$$\forall t \in [T_{d,b}, T_{d,e}], P_t^{EV,d} \leq P^{EV,d,rated} \tag{13}$$

$$\varnothing = [T_{c,b}, T_{c,e}] \cap [T_{d,b}, T_{d,e}] \tag{14}$$

Equations (7) and (8) indicate the state relations between the initial EV energy of the first day ($W^{EV,(1)}$), the EV remaining energy ($W^{EV,rem}$), and the energy consumption on the daily trip ($W^{EV,trip}$). In addition, Eq. (9) describes that the remaining energy of EV can be used to cover a portion of energy usage by household appliances via battery discharging ($W^{EV,d}$) and the EV will be charged to an appropriate level for the usage of the second day.

Moreover, Eq. (10) explains the relationship between the total energy charging ($W^{EV,c}$) and the charging power rate ($P_t^{EV,c}$). Parameter η_1 denotes the battery

Charging circuit	Charging power (kW)	Charging speed (miles/hour)	Time cost per 100 miles (hour)
Wall connector (one-phase grid)	7.4	22	4.5
Wall connector (three-phase grid)	11	34	2.9
High-power charger upgrade	16.5	51	2.0
Three-pin domestic adapter	2.3	6.8	14.7

Table 2.
Tesla Model S charging schemes.

charging efficiency. Time parameters $T_{\text{c, b}}$ and $T_{\text{c, e}}$ represent the begin time and the end time of the charging operation. Meanwhile, the meanings of variables of the battery discharging occasion, which is described in Eq. (11), are similar to those in Eq. (10).

Further, constraint in Eq. (12) presents a limit on the actual amount energy of the EV battery. It cannot drop below the minimum allowed battery capacity ($W^{\text{EV, min}}$) or exceed the maximum allowed battery capacity ($W^{\text{EV, max}}$). Constraint in Eq. (13) limits the actual discharging power rate ($P_t^{\text{EV, d}}$) to be less than the rated power of the EV. Additionally, since battery damages will be caused by the simultaneous charging and discharging, constraint in Eq. (14) restricts the operation time of battery charging and discharging.

2.3 Problem formulation and optimization

According to the previous analysis, the problem in this study can be formulated as minimizing the total cost (TC) by scheduling the operating time of the household appliances. Hence, the objective function can be proposed as

$$\text{Min TC} = \int_{T_{\text{in}}}^{T_{\text{term}}} W_t^{\text{grid}} \cdot R_t \cdot d(t) \tag{15}$$

where the variable W_t^{grid} represents the total energy bought from the power grid in time period $[T_{\text{in}}, T_{\text{term}}]$. Additionally, the price variable R_t is time dependent and varies hourly depending on the total load demand [10]. The DP tariff that is used in simulations is given in **Figure 3** in Case study and results section.

In order to obtain the optimal solution and reduce the cost to the minimum, the exhaustive search technique can be used on the basis of the established models. The detail description of the technique is not the focus of this work, so it is not emphasized here.

Note that the remaining EV energy is suggested to be firstly consumed in high-price hours to ensure the maximum electric bill reduction. The feasibility of the proposed EV-APS DR strategy is evaluated in Case 1 in Section 4.

3. Electrical vehicle-assisted demand response strategy for multi-household network

Based on the achievement of Section 2, this section proposes a demand response strategy with multiple EVs for a multi-household network. An EV-assisted DR strategy with a neighbor energy sharing (NES) model is described first. After that, the system models are introduced mathematically in details. At last, the problem formulation and optimization are illustrated.

3.1 EV-APS demand response network

The block diagram of the proposed DR framework with the EV-assisted NES model is shown in **Figure 2**. In this study, it is assumed that each household in the community is registered in the network and controlled by the corresponding automatic control unit (ACU) which plays the role as an instructor of each household. ACU regulates the power supplying and the operating time of the household appliances (HAs, e.g., flexible appliances and critical appliance) based on the dynamic load information which is usually received from smart meters and other request

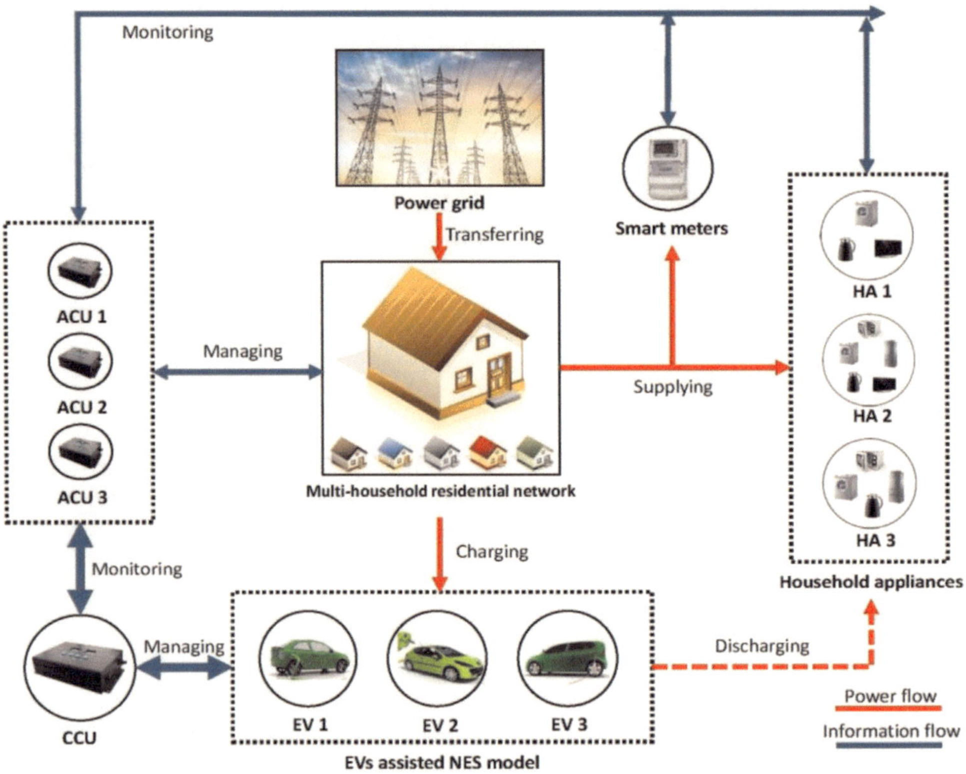

Figure 2.
Schematic diagram of a NES model-based DR strategy for a multi-household network.

signals (e.g., EV status, scheduling priority, DP, etc.). In addition, the centralized control unit (CCU) that is the highest controller in the network globally monitors the status of the ACUs and optimally manages the EV-assisted NES model through the information flows. In the proposed DR framework, customers in the network are registered for two types of connections: V2H connection and V2N connection.

Specifically, the householders buy electricity from the power grid for the daily consumption including HA supplying and EV charging, under the DP tariff. On the one hand, the domestic appliances are directly powered by the public power grid in general. However, the household which is outfitted with EV is able to provide power from EV battery for their HAs on appropriate occasions, such as peak demand periods or power grid outage, via V2H connection. On the other hand, since a limited number of the households are equipped with EV at their premises, the households without energy storage unit may need power assistance from NES model via V2N connection, particularly in high-price periods. When there is surplus energy available being detected in EVs, the CCU determines when and how to allocate the surplus energy to the personal house or the neighbor's houses who have the energy assistance requirements. Generally, the EV energy will satisfy the demand of the EV owner in priority. The energy transaction in neighborhood happens when the power grid is not able to fulfill the demand or the serving load at high charges in peak demand periods. Thus, a customer can receive the power from a neighbor at comparatively lower prices.

The mathematical models of the proposed DR framework will be discussed in the next subsection.

3.2 System models

EVs are utilized as the flexible energy storage units to ensure the energy trading in neighborhood. The following subsections present the mathematical modeling of the system components in details.

3.2.1 Global energy balance model

In order to precisely present the energy transactions between each component in the network with K households, W_t^{grid} and $W_{k,t}^{\text{grid}}$ are defined as the total energy consumption of the entire network and the kth household, respectively, in a time period $[T_{\text{in}}, T_{\text{term}}]$. Afterwards, the global energy model can be proposed as in Eq. (16):

$$W_t^{\text{grid}} = \sum_{k=1}^{K} W_{k,t}^{\text{grid}} \tag{16}$$

where

$$W_{k,t}^{\text{grid}} = \int_{T_{\text{in}}}^{T_{\text{term}}} P_{k,t}^{\text{grid}} \cdot d(t) \tag{17}$$

Moreover, considering the specific power including CS appliances ($P_{k,t}^{\text{CS}}$) and FS appliances ($P_{k,t}^{\text{FS}}$), EV charging ($P_{k,t}^{\text{EV, c}}$), and EV discharging ($P_{k,t}^{\text{EV, d}}$) into the network, $P_{k,t}^{\text{grid}}$ in Eq. (17) can be extended as in Eqs. (18) and (19):

$$P_{k,t}^{\text{grid}} = P_{k,t}^{\text{HA}} + \alpha \cdot \left(\beta \cdot P_{k,t}^{\text{EV, c}} - (1-\beta) \cdot P_{k,t}^{\text{EV, d}}\right) \tag{18}$$

$$P_{k,t}^{\text{HA}} = \sum_{j=1}^{m} P_{k,t,j}^{\text{CS}} + \varepsilon_i \sum_{i=1}^{n} P_{k,t,i}^{\text{FS}} \tag{19}$$

Subject to

$$\forall t, P_{k,t}^{\text{grid}} \le P_{k,\max}^{\text{grid}} \tag{20}$$

$$\forall t, \sum_{k=1}^{K} P_{k,t}^{\text{grid}} \le P_{\max}^{\text{grid}} \tag{21}$$

Binary parameters α and β in Eq. (18) are both used to indicate the EV status that is given as

$$\text{EV status} = \begin{cases} \text{Disabled,} & \text{if } \alpha = 0, \beta = \forall \\ \text{Charging,} & \text{if } \alpha = 1, \beta = 1 \\ \text{Discharging,} & \text{if } \alpha = 1, \beta = 0. \end{cases}$$

Furthermore, $P_{k,t}^{\text{HA}}$ in Eq. (18) denotes the load of electrical appliances consisting of CS load $P_{k,t,j}^{\text{CS}}$ and FS load $P_{k,t,j}^{\text{FS}}$ at time t, where j and i represent the index of the appliances. The ε parameter indicates the scheduling priorities of the scheduled appliances, which is similar to Eq. (3). Besides, the maximum power rate of an individual household $P_{k,\max}^{\text{grid}}$ and the maximum power rate of the network $P_{\max}^{\text{grid}}$ are proposed in Eqs. (19) and (20), respectively, to limit the real-time load for the safety consideration.

3.2.2 EV-assisted NES model

In a residential community, different classes of customers exist. It is not possible for every household to purchase an EV. Thus, it is assumed that only a part of houses are installed with EV and indexed as $\hat{k}$, and the rest houses without EV are indexed as $\tilde{k}$. Similar to the EV-APS model, we define $W_{\hat{k}}^{\text{EV},(1)}$ and $W_{\hat{k}}^{\text{EV},(2)}$ as the initial energy within the kth EV battery when EV leaves home of the first day and second day, respectively. Variable $W_{\hat{k}}^{\text{EV,rem}}$ represents the remaining energy within the kth EV. The energy cost of the kth EV on the daily trip is proposed as $W_{\hat{k}}^{\text{EV,trip}}$. Additionally, $D_{\hat{k}}^{\text{trip}}$ and $D_{\hat{k}}^{\max}$ are proposed to indicate the actual travel distance of vehicle and the maximum travel distance with a fully charged EV. Moreover, the energy charging to EV and discharging from EV are assumed as $W_{\hat{k}}^{\text{EV,c}}$ and $W_{\hat{k}}^{\text{EV,d}}$, respectively. Afterwards, the EV balance model with the relevant constraints for the kth EV household can be proposed as follows:

$$W_{\hat{k}}^{\text{EV,rem}} = W_{\hat{k}}^{\text{EV},(1)} - W_{\hat{k}}^{\text{EV,trip}} \tag{22}$$

$$W_{\hat{k}}^{\text{EV,trip}} = \frac{D_{\hat{k}}^{\text{trip}}}{D_{\hat{k}}^{\max}} \cdot W_{\hat{k}}^{\text{EV,max}} \tag{23}$$

$$W_{\hat{k}}^{\text{EV},(2)} = W_{\hat{k}}^{\text{EV,rem}} + W_{\hat{k}}^{\text{EV,c}} - W_{\hat{k}}^{\text{EV,d}} \tag{24}$$

Subject to

$$\forall t,\ W_{\hat{k}}^{\text{EV,min}} \leq W_{\hat{k}}^{\text{EV,rem}} \leq W_{\hat{k}}^{\text{EV,max}} \tag{25}$$

$$\tau \cdot W_{\hat{k}}^{\text{EV,max}} \leq W_{\hat{k}}^{\text{EV},(1)} \approx W_{\hat{k}}^{\text{EV},(2)} \leq W_{\hat{k}}^{\text{EV,max}} \tag{26}$$

where variables $W_{\hat{k}}^{\text{EV,min}}$ and $W_{\hat{k}}^{\text{EV,max}}$ in Eq. (25) represent the minimum and the maximum allowed EV battery capacity, respectively. However, constraint in Eq. (26) is proposed to ensure the EV leaves home with an appropriate energy storage level, where τ is a threshold parameter.

Moreover, considering the power impact in the multi-household network, $P_{\hat{k},t}^{\text{EV,c}}$, $P_{\hat{k},t}^{\text{EV,d,v2h}}$, and $P_{\hat{k},t}^{\text{EV,d,v2n}}$ are utilized to describe the power rates of EV charging, EV discharging via V2H, and EV discharging via V2N at time t, respectively. Therefore, $W_{\hat{k}}^{\text{EV,c}}$ and $W_{\hat{k}}^{\text{EV,d}}$ in Eq. (24) can be extended as

$$W_{\hat{k}}^{\text{EV,c}} = \eta_{\hat{k}}^{\text{c}} \cdot \left\{ \sum_{l=1}^{L} \int_{T_{\hat{k},l}^{\text{c},1}}^{T_{\hat{k},l}^{\text{c},2}} P_{\hat{k},t}^{\text{EV,c}} \cdot d(t) \right\} \tag{27}$$

$$W_{\hat{k}}^{\mathrm{EV,\,d}} = \frac{1}{\eta_{\hat{k}}^{\mathrm{d,\,v2h}}} \cdot \left\{ \sum_{m=1}^{M} \int_{T_{\hat{k},m}^{\mathrm{d},1}}^{T_{\hat{k},m}^{\mathrm{d},2}} P_{\hat{k},t}^{\mathrm{EV,\,d,\,v2h}} \cdot d(t) \right\} + \frac{1}{\eta_{\hat{k}}^{\mathrm{d,\,v2n}}} \cdot \left\{ \sum_{n=1}^{N} \int_{T_{\hat{k},n}^{\mathrm{d},1}}^{T_{\hat{k},n}^{\mathrm{d},2}} P_{\hat{k},t}^{\mathrm{EV,\,d,\,v2n}} \cdot d(t) \right\} \tag{28}$$

Subject to

$$\eta_{\hat{k}}^{\mathrm{c}}, \eta_{\hat{k}}^{\mathrm{d,\,v2h}} \text{ and } \eta_{\hat{k}}^{\mathrm{d,\,v2n}} \in (0, 1) \tag{29}$$

$$\forall t \in \left[T_{\hat{k},m}^{\mathrm{d},1}, T_{\hat{k},m}^{\mathrm{d},2}\right], P_{\hat{k},t}^{\mathrm{EV,\,d,\,v2h}} \le P_{\hat{k}}^{\mathrm{EV,\,rated}}, P_{\hat{k},t}^{\mathrm{EV,\,d,\,v2h}} \le P_{\hat{k},t}^{\mathrm{act}} \tag{30}$$

$$\forall t \in \left[T_{\hat{k}}, n^{\mathrm{d},1}, T_{\hat{k}}, n^{\mathrm{d},2}\right], P_{\hat{k},t}^{\mathrm{EV,\,d,\,v2n}} \le P_{\hat{k}}^{\mathrm{EV,\,rated}}, P_{\hat{k},t}^{\mathrm{EV,\,d,\,v2n}} \le P_{\tilde{k},t}^{\mathrm{act}} \tag{31}$$

$$\varnothing = \forall \left[T_{\hat{k},l}^{\mathrm{c},1}, T_{\hat{k},l}^{\mathrm{c},2}\right] \cap \forall \left\{ \left[T_{\hat{k},m}^{\mathrm{d},1}, T_{\hat{k},m}^{\mathrm{d},2}\right] \cup \left[T_{\hat{k},n}^{\mathrm{d},1}, T_{\hat{k},n}^{\mathrm{d},2}\right] \right\} \tag{32}$$

where $\eta_{\hat{k}}^{\mathrm{c}}$, $\eta_{\hat{k}}^{\mathrm{d,\,v2h}}$, and $\eta_{\hat{k}}^{\mathrm{d,\,v2n}}$ denote the efficiencies of the corresponding EV behaviors. Since the EV behaviors are discontinuous and may execute at different periods, different time labels are proposed. For example, time parameters $T_{\hat{k},l}^{\mathrm{c},1}$ and $T_{\hat{k},l}^{\mathrm{c},2}$ in Eq. (26) represent the start time and the end time of lth charging period. The definitions of the time parameters in EV discharging periods as shown in Eq. (27) are similar to Eq. (26).

Furthermore, the discharging power via V2H connection ($P_{\hat{k},t}^{\mathrm{EV,\,d,\,v2h}}$) cannot exceed the rated power ($P_{\hat{k}}^{\mathrm{EV,\,rated}}$) nor the actual power required of the household ($P_{\hat{k},t}^{\mathrm{act}}$) as shown in Eq. (30). Constraint in Eq. (31) is similar to (30), which limits the discharging power via V2H connection ($P_{\hat{k},t}^{\mathrm{EV,\,d,\,v2n}}$). Variable $P_{\tilde{k},t}^{\mathrm{act}}$ in Eq. (31) represents the actual load demand of the neighbor which receives the power assistance from the EV household via V2N connection. Besides, as shown in Eq. (32), the EV charging and discharging are not allowed to operate simultaneously as well for the purpose of protecting the EV battery from damage.

3.2.3 Energy trading model in neighborhood

The proposed EV-assisted NES model ensures the energy trading in neighborhood via V2H and V2N connections. However, it is necessary to declare the trading policy in neighborhood in advance, which is illustrated as follows:

1. The EV energy will be provided in priority to satisfy the load demand of the household which owns the EV.

2. After (1), the surplus EV energy will be used in priority to supply the households which are not equipped with any energy storage units (e.g., EVs.).

3. If multiple EVs have surplus energy, the EV with the most energy reserve will be adopted in priority to assist neighbors' load demand.

4. If multiple households require energy assistance, the household which requires more load demand during high-price period will receive the energy sharing in

priority, and each house can obtain energy assistance from only one EV energy provider.

5. The allocation of the EV energy will follow the principle of maximizing the benefits of the EV provider.

In addition to these, $B_{\hat{k}}^{\mathrm{NES}}$ and $B_{\tilde{k}}^{\mathrm{NES}}$ are proposed to describe the obtained benefit of the households who sold EV energy and received energy assistance, respectively, via NES model. Hence, $B_{\hat{k}}^{\mathrm{NES}}$ and $B_{\tilde{k}}^{\mathrm{NES}}$ can be formulated as follows:

$$B_{\hat{k}}^{\mathrm{NES}} = \theta\% \cdot \left(C_{\tilde{k}}^{\mathrm{dmd}} - C_{\hat{k}}^{\mathrm{EV,\,c}}\right) \tag{33}$$

$$B_{\tilde{k}}^{\mathrm{NES}} = (1 - \theta\%) \cdot \left(C_{\tilde{k}}^{\mathrm{dmd}} - C_{\hat{k}}^{\mathrm{EV,\,c}}\right) \tag{34}$$

Subject to

$$C_{\tilde{k}}^{\mathrm{dmd}} - C_{\hat{k}}^{\mathrm{EV,\,c}} > 0 \tag{35}$$

where θ is a profit distribution parameter and normally $\theta\% = 0.5$, which means the participants in energy trading share the profits equally. Additionally, $C_{\tilde{k}}^{\mathrm{dmd}}$ is the cost for electricity demand without EV sharing within household without EV equipment, and $C_{\hat{k}}^{\mathrm{EV,\,c}}$ is the cost for EV charging of the energy sharing part. However, the energy transaction via NES model occurs only when it is profitable as shown in Eq. (35). Obviously, this type of EV-based energy sharing model is benefit for the trading participants on both sides.

3.3 Problem formulation and optimization

The objective of this work is to minimize the total daily cost for energy usage of the residential network with K households as well as shape the load to a proper level in peak demand time. To begin with, the day is split into equal time divisions with a time interval and indexed as t. The total cost function is given in Eq. (35):

$$\text{Min TC} = \sum_{k=1}^{K} \left\{ \sum_{t=1}^{24} \left(R_t \cdot W_{k,t}^{\mathrm{grid}}\right) - B_k^{\mathrm{NES}} \right\} \tag{36}$$

where R_t is the dynamic electricity pricing, $W_{k,t}^{\mathrm{grid}}$ is the energy consumed on the grid of the kth household, and B_k^{NES} represents the cost benefit that the householder can obtain in energy trading in neighborhood by using the proposed NES model.

According to the defined trading policies between neighbors, in order to minimize TC, we have to minimize each TC_k which denotes the total cost of the kth household in the network. Therefore, the objective function can be formulated as

$$\text{Min TC} = \sum_{k=1}^{K} \text{Min}\left\{\mathrm{TC}_k\right\} = \sum_{k=1}^{K} \left\{ \text{Min}\left\{ \sum_{t=1}^{24} \left(R_t \cdot W_{k,t}^{\mathrm{grid}}\right) \right\} - \text{Max}\left\{B_k^{\mathrm{NES}}\right\} \right\} \tag{37}$$

Based on the objective function in Eq. (36), the optimization process can be executed in two stages. First, it minimizes the total cost for electricity bill by optimally allocating the EV energy via V2H connection. Second, it maximizes the benefit from energy transaction in neighborhood by optimally distributing the surplus energy via V2N connection. Based on the previous model descriptions, both optimization stages are linear problems. Therefore, the mixed-integer linear programming (MILP) which is the most appropriate technique has been used to obtain the optimal solution. However, the description of the technique is not the focus in this study, so it is not emphasized here.

Under the given models and the relevant constraints, the proposed DR strategy is able to optimally schedule appliances within the multi-household network in accordance with the comprehensive affecting factors, such as EV behaviors, user preferences, and load scheduling priorities. Here, the maintenance cost for EVs and home appliances is neglected in this work.

4. Case study and results

In order to evaluate the feasibility of the proposed DR strategies, two cases are proposed in this section. Case 1 is used to evaluate the EV-assisted DR strategy for an individual household, and Case 2 is proposed to evaluate the EV-assisted DR strategy for a multi-household network with multiple EVs.

4.1 Case 1: EV-assisted DR strategy for individual household

This subsection demonstrates how the proposed EV-APS DR strategy can be implemented at the household level to alleviate the load burden in peak demand periods and save electric bills. Some assumptions for simulations are presented.

4.1.1 Case description

In this case, the selected time interval for the optimization is set as 3 minutes (0.05 hr). The households comprise over 15 types of commonly used loads covering both CS and FS appliances. The EV and four other commonly used appliances, hot water tank, dish machine, washer, and drying machine, are considered as the flexible loads in this study.

In addition, the ε parameters are given to indicate the priorities of the related loads. According to the user preferences, it is randomly assumed. Besides, in accordance with the operating habits, the objective scheduling time for these appliances is set randomly, such as EV charging [0:00–8:00]; hot water tank [17:00–22:00]; dish machine [18:30–24:00]; washer [17:00–24:00]; and drying machine [0:00–8:00].

Moreover, the Tesla Model S (EV) with a battery rating of 30 kWh (up to 60 kWh) is employed in the case study. It is provided with a charging wall connector (one-phase grid) limited to a charging power of 7.4 kW. The discharging power for household appliances is up to 3.0 kW as it is shown in **Table 2**. The charging and discharging efficiencies are considered as $\eta_1 = \eta_2 = 0.95$. It is also considered that the householder always arrives home at 5:00 p.m. with 18 kWh (60%) remaining energy in EV battery and leaves home at 8:00 a.m. in the next morning with fully charged battery (100 $\pm$ 5%, 30 $\pm$ 1.5 kWh). However, the minimum remaining energy in EV is restricted to 7.5 kWh (25 $\pm$ 5%) to avoid the deep discharging. The deep charging will cause damages to the battery and reduce battery life. Furthermore, the UK dynamic pricing data of a typical day which is used in this case is presented in **Figure 3**.

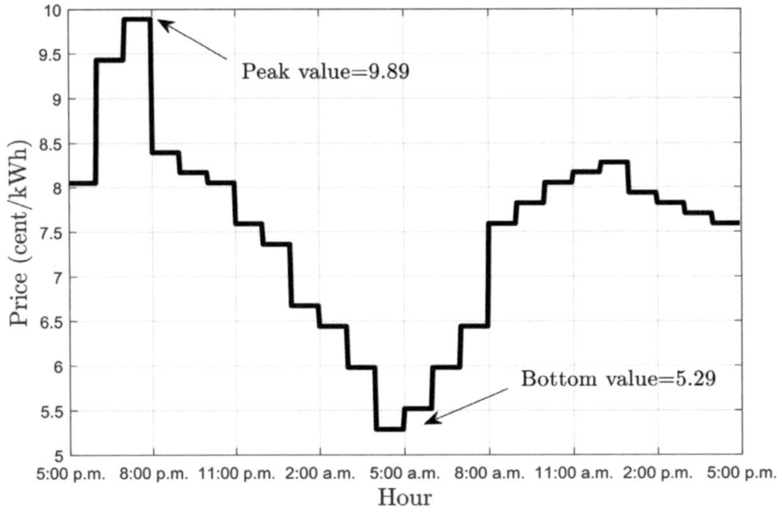

Figure 3.
UK real-time electricity pricing data.

4.1.2 Simulation result

Assuming that the target household demand limits of 8 kW all day in this study, **Figure 4** presents the overall load shaping results of the household appliances. Specifically, **Figure 4(a)** shows the original load profile without DR. It can be seen that the peak demand time occurs between 6:00 p.m. and 8:10 p.m. The total house load exceeds the 8 kW limit during this period, and the maximum load demand is 11.5 kW which occurs at around 8:00 p.m. Additionally, (b) and (c) in **Figure 4** present the load profiles after scheduling by using the LSC DR strategy [20] and the proposed EV-APS DR strategy, respectively. Apparently, the load burden is alleviated, and the load decreases to an appropriate level in both (b) and (c). Nonetheless, compared with the results in (b), the load demand in (c) between 6:00 p.m. and 9:40 p.m. approaches to a very low level, since the EV discharging is activated during this time. As a consequence, the EV takes 3.2 hour to charge as it is shown in (c), which is longer than the charging time (2.1 hour) in (b).

Moreover, since the EV plays a great role in power supplying in modeling, the real-time EV remaining energy variation at household parking station by using the proposed EV-APS DR strategy is illustrated in **Figure 5**. Specifically, the EV arrives at home at 5:00 p.m. as described in the figure. Between 5:00 p.m. and 10:18 p.m., the EV discharging is activated, and a part of household appliances are continuously powered by EV until the amount of EV remaining energy reaches the minimum threshold (7.5 kWh). However, the EV is charged from 3:00 a.m. to 6:18 a.m. in the next day morning to enable the EV leaves home with the fully charged battery at 8:00 a.m. According to the results, it can be seen that the EV remaining energy variation directly corresponds with the load curve in **Figure 4(c)**, which indicates that this emulation method is correct and feasible.

Figure 6 shows the accumulative probabilities of the reshaped load distributions by DR strategies during peak load demand period which is between 5:00 p.m. and 12:00 p.m. Based on the figure, we can see that the probabilities for the case $P^{grid} <$ 1 kW of the original load profile without DR, the LSC DR shaping profile, and the EV-APS DR shaping profile are 7.1%, 24.3% and 72.9%, respectively. For the case $P^{grid} <$ 3 kW, the probabilities are 23.6, 53.1 and 86.4%, respectively. The results indicate that the load shaping performance by the EV-APS DR strategy is the best as

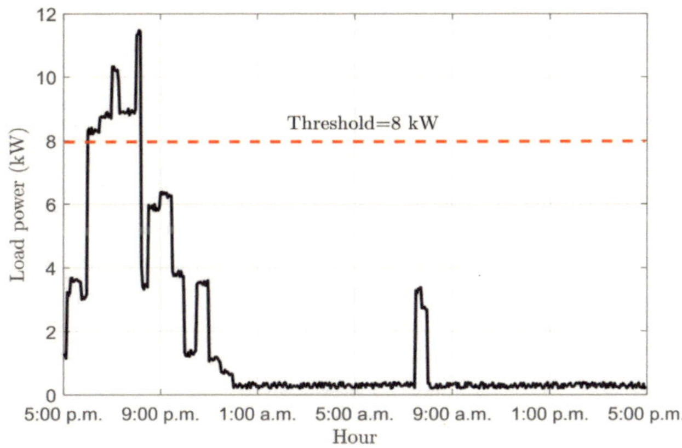

(a)

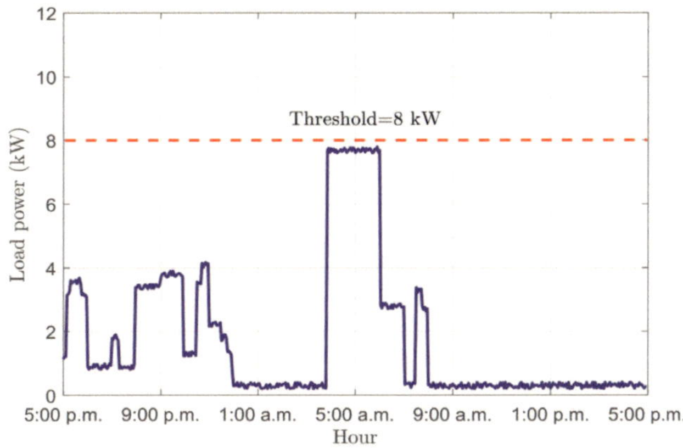

(b)

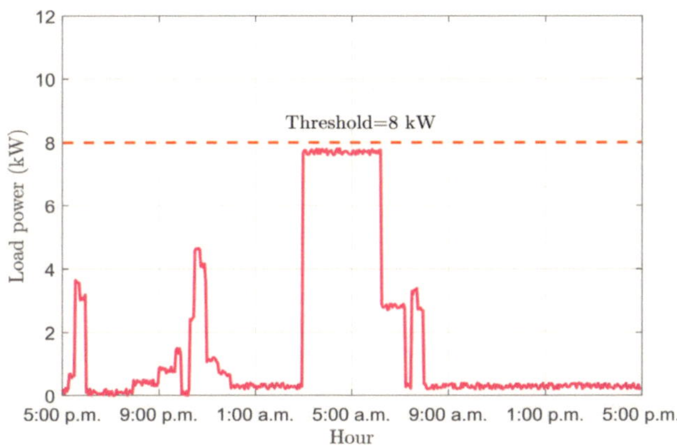

(c)

Figure 4.
The overall load shaping results. The load profiles of (a) without DR, (b) by the LSC DR, and (c) by the proposed EV-APS DR.

a higher percentage load is shaped to a low level, which proves that the proposed method is an effective tool in load shaping.

The total cost is another issue that customers concern. On the basis of the DP tariff, the daily electric cost can be obtained. **Figure 7** presents the accumulative

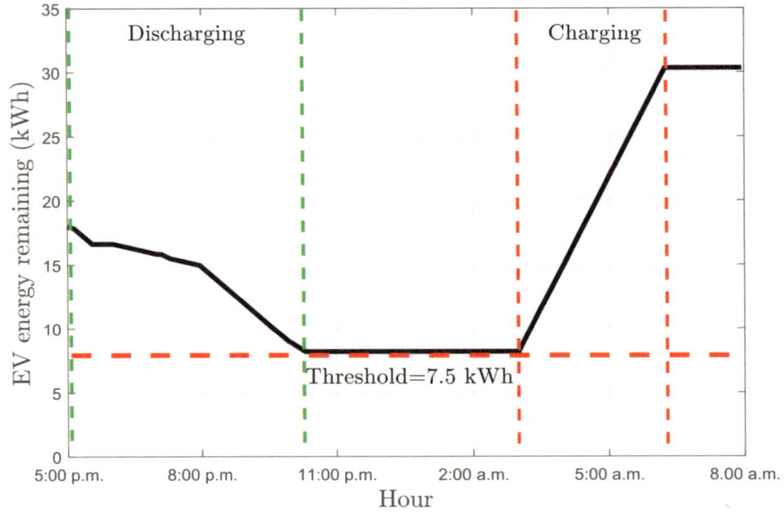

Figure 5.
The real-time EV remaining energy variation at parking station.

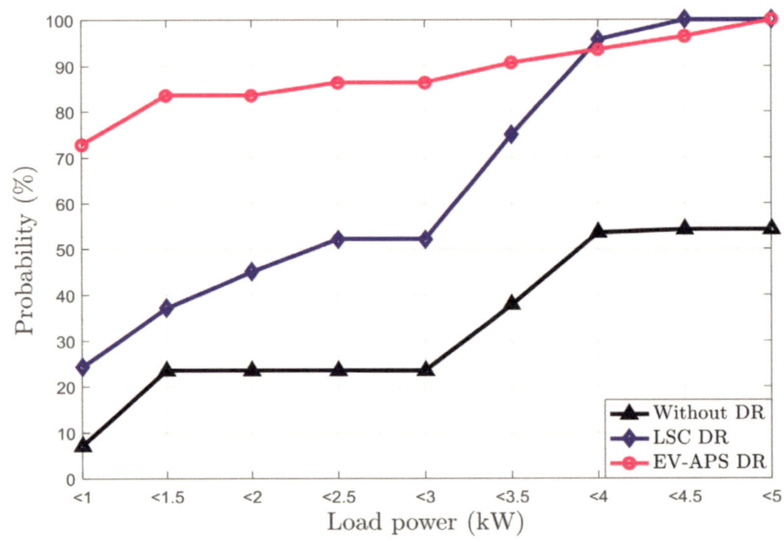

Figure 6.
The accumulative probability of the load distribution during peak load demand hours.

cost comparison between different demand response strategies. Obviously, the proposed EV-APS DR strategy performs superior than other approaches in comparison. The total electric bill of the original load demand of a typical day is about £3.6. However, it decreases to £2.9 and £2.5 by using the LSC DR and the EV-APS DR, respectively. The total saving costs are about £0.7 and £1.1, which are equivalent to 19.4 and 30.6%, respectively. Compared with the LSC DR strategy in literature, the proposed DR strategy in this paper has a better performance in load shaping and higher cost saving percentage (11.2% improved), obviously.

4.2 Case 2: EV-assisted DR strategy for multi-household network

This section proposes a case study to demonstrate how the DR strategy with the EV-assisted NES model can be implemented at the side of residential community, to

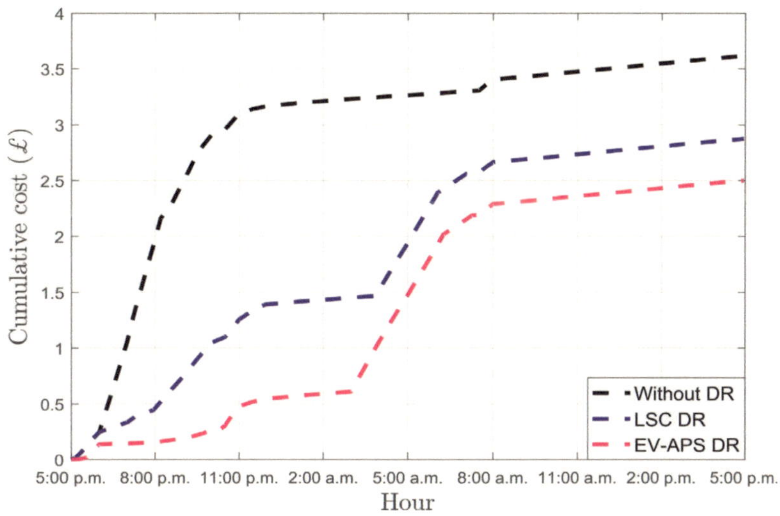

Figure 7.
The accumulative cost comparison results between DR strategies.

Parameter	House #1	House #2	House #3	House #4	House #5
EV status	Active	Active	Active	Disable	Disable
ToA (first day)	5 p.m.	6 p.m.	7 p.m.	—	—
ToL (second day)	8 a.m.	9 a.m.	10 a.m.	—	—
CR (kW)	7.5	6.5	5.5	—	—
DCR (kW)	3.5	3	2.5	—	—
ERoA (kWh)	26	24	22	—	—

Table 3.
Electrical vehicle parameter specification.

save electric bills and alleviate the load burden in peak demand time simultaneously.

4.2.1 Case description

The optimization problem for the total cost minimization is formulated as linear programming aimed to reduce the daily bill of each household as much as possible. In the case study, the selected time interval for the optimization is set as 3 minutes. The adopted multi-household network is assumed to comprise five households for convenience. For each household, over 15 types of commonly used domestic appliances covering both FS and CS appliances are accounted.

In addition, as not all the users are able to purchase an electrical vehicle, only 3/5 of the households are assumed to be equipped with EVs to support the neighbor energy sharing. For each EV device, a battery capacity of 35 kWh is employed. The charging and discharging (via V2H and V2G) efficiencies are all considered to be 0.95 for convenience. The minimum remaining energy in EV is restricted to 10% ($\tau = 0.1$) of the battery capacity to avoid the deep discharging.

Besides, the parameters about the EV status, time of arriving (ToA), time of leaving (ToL), charging rate (CR), discharging rate (DCR), and energy remaining of arriving home (ERoA) of the specific EV within each household are given in **Table 3**.

4.2.2 Simulation results

Figure 8 presents the overall load shaping results for the multi-household network by using different DR programs. It is assumed that the threshold of the overall load demand is 25 kW. Specifically, it can be seen that the LSC demand response strategy can slightly alleviate the load burden, particularly around 9 p.m. This is because limited appliances are scheduled, and none of EVs are adopted in the LSC DR program. However, the load shaping performances of using EVs without NES and EV-assisted NES in (c) and (d), respectively, are much better than the results in (a) and (b). The load demand of the entire network in both (c) and (d) has remained below the threshold apparently due to the EV discharging contributions.

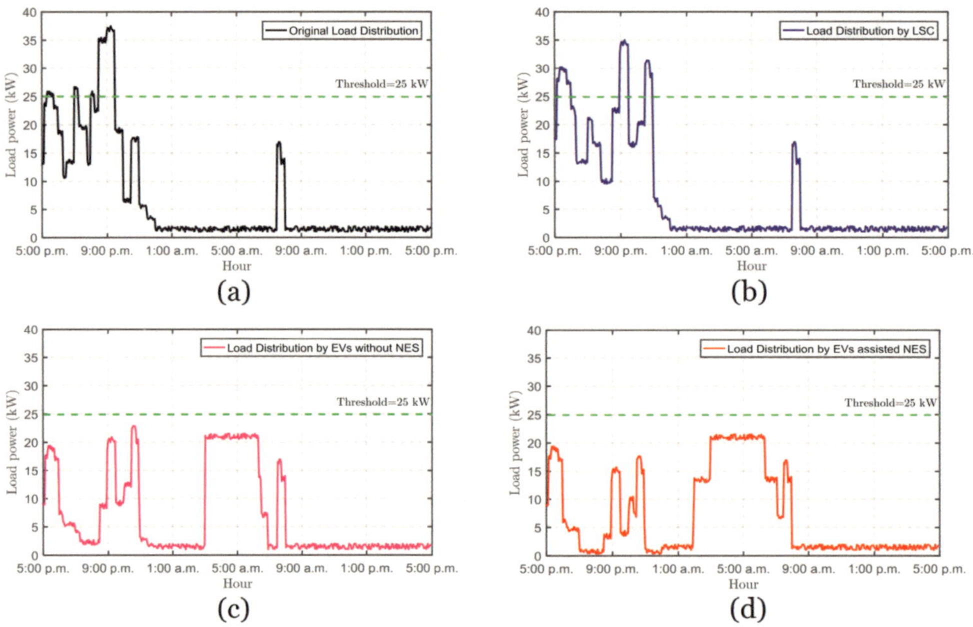

Figure 8.
Overall load shaping results for the multi-household network by using different DR programs. The load profiles of (a) without DR, (b) by LSC DR, (c) by EV without NES DR, and (d) by EV-assisted NES DR.

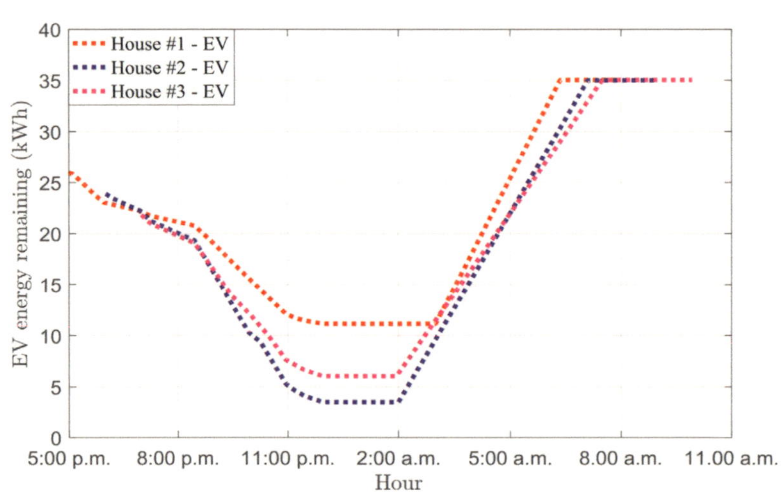

Figure 9.
Real-time energy remaining variations of EVs at parking station.

Methods	House #1	House #2	House #3	House #4	House #5
Original	3.15	3.55	3.88	2.31	2.34
LSC	3.09	3.49.	3.73	2.24	2.26
EVs without NES	1.83	2.09	2.30	—	—
EVs with NES	1.83	1.87	2.13	2.07	2.02

Table 4.
Daily cost (£) comparison by adopting different DR programs.

In addition, compared with the load distribution in (c), the load demand in (d) approaches to a lower level in peak time around 7–9 p.m. This is because the households without EVs received the energy assistance from neighbors via V2N so that the overall load demand on the grid decreases. As a consequence, it is obvious to see that the EVs take more time to charge the batteries in off-peak time for the usage of the second day. Moreover, since the EVs play a great role in power transaction within the network, the real-time energy remaining variations of EVs (#1, #2, and #3) at parking station are illustrated in **Figure 9**.

In terms of the daily electricity cost, the proposed approach can obtain more benefits compared with the literature DR programs as shown in **Table 4**. According to the cost results, apparently, the proposed DR with an EV-assisted NES model performs the best with the lowest cost in the comparison for all cases. Specifically, as house #1 does not participate in the energy trading in neighborhood due to the lower distributing priority, there is no cost difference between using EVs with and without NES. Nonetheless, as the energy providers in the transaction, the costs of house #2 and house #3 are reduced by 47.3 and 46.1%, respectively, by adopting the EV-assisted NES model compared with the original cost. Additionally, about 10.5 and 7.4% cost reduction can be achieved compared with the method of using EVs without NES. On the other side of the trading, house #4 and house #5 that are not equipped with EVs also obtain the benefits from the energy sharing. About £0.24 and £0.32 which are equivalent to 10.4% and 13.7% cost saving can be gained during the transaction for house #4 and house #5, respectively.

In overview, for this selected residential community including five individual households, the total payment saving is about £5.31 which is equivalent to 34.9% in this case. Obviously, the adopted EV-based NES model is beneficial for the energy trading participants on both sides, and significant improvements can be achieved comparing with the literature DR programs.

5. Conclusion

The aim of this work is to develop DR strategies assisted by EVs, to jointly optimize the household appliance scheduling and economic cost based on DP for different scales of households. An EV-APS-based DR strategy has been proposed first and then extended to an EV-NES model-based DR strategy. The numerical results demonstrated that for by using the EV-APS-based DR strategy for a single household, 86.4% of the load in peak hours can be shifted to an off-peak time and that the daily electric cost can be reduced by 30.6%. For the multi-household network, the load can be significantly shifted to an appropriate level, and the daily electric cost of the entire network can be reduced by 34.9%. On the basis of the achieved results, we can conclude that the proposed DR strategies in this chapter are energy-efficient solutions and can fulfill the tasks of load balancing and cost saving for the smart grid and customers simultaneously.

Acknowledgements

A part of the chapter was taken from the paper entitled "Dynamic pricing based and electric vehicle assisted demand response strategy," which has been published in 2017 IEEE International Conference on Smart Grid Communications (SmartGridComm), and we have obtained the permission to reuse it.

Conflict of interest

The authors declare that they have no conflicts of interest.

Funding

This work is partially supported by the XJTLU Research Development Fund (PGRS-13-03-06, RDF-14-03-24 and RDF-14-02-48) and AI University Research Centre (AI-URC) through XJTLU Key Program Special Fund (KSF-P-02).

Author details

Xing Luo[1,2], Xu Zhu[1,3]* and Eng Gee Lim[2]

1 Department of Electrical Engineering and Electronics, University of Liverpool, UK

2 Department of Electrical and Electronic Engineering, Xi'an Jiaotong-Liverpool University, P. R. China

3 School of Electronic and Information Engineering, Harbin Institute of Technology, Shenzhen, P. R. China

*Address all correspondence to: xuzhu@liverpool.ac.uk

References

[1] Yao L, Lim WH, Tsai TS. A real-time charging scheme for demand response in electric vehicle parking station. IEEE Transactions on Smart Grid. 2017;**8**(1): 52-62. DOI: 10.1109/TSG.2016.2582749

[2] Chen Q, Wang F, Hodge BM, Zhang J, Li Z, Shafie-Khah M, et al. Dynamic price vector formation model-based automatic demand response strategy for PV-assisted EV charging stations. IEEE Transactions on Smart Grid. 2017;**8**(6): 2903-2915. DOI: 10.1109/TSG.2017. 2693121

[3] Brooks A, Lu E, Reicher D, Spirakis C, Weihl B. Demand dispatch. IEEE Power and Energy Magazine. 2010;**8**(3): 20-29. DOI: 10.1109/MPE.2010.936349

[4] O'Dwyer C, Duignan R, O'Malley M. Modeling demand response in the residential sector for the provision of reserves. In: Proceedings of IEEE Power and Energy Society General Meeting; 22-26 July 2012; San Diego, USA. pp. 1-8

[5] Ansari M, Al-Awami AT, Sortomme E, Abido MA. Coordinated bidding of ancillary services for vehicle-to-grid using fuzzy optimization. IEEE Transactions on Smart Grid. 2015;**6**(1): 261-270. DOI: 10.1109/TSG.2014. 2341625

[6] Yilmaz M, Krein PT. Review of the impact of vehicle-to-grid technologies on distribution systems and utility interfaces. IEEE Transactions on Power Electronics. 2013;**28**(12):5673-5689. DOI: 10.1109/TPEL.2012.2227500

[7] Shao S, Pipattanasomporn M, Rahman S. Grid integration of electric vehicles and demand response with customer choice. IEEE Transactions on Smart Grid. 2012;**3**(1):543-550. DOI: 10.1109/TSG.2011.2164949

[8] Yoon SG, Choi YJ, Park JK, Bahk S. Stackelberg-game-based demand response for at-home electric vehicle charging. IEEE Transactions on Vehicular Technology. 2016;**65**(6): 4172-4184. DOI: 10.1109/ TVT.2015.2440471

[9] Ferreira JC, Monteiro V, Afonso JL. Vehicle-to-anything application for electric vehicles. IEEE Transactions on Industrial Informatics. 2014;**10**(3): 1927-1937. DOI: 10.1109/TII.2013. 2291321

[10] Lo KL, Wu YK. Analysis of relationships between hourly electricity price and load in deregulated real-time power markets. IEE Proceedings— Generation, Transmission and Distribution. 2004;**151**(4):441-452. DOI: 10.1049/ip-gtd: 20040613

[11] Zhao J, Kucuksari S, Mazhari E, Son Y-J. Integrated analysis of high-penetration pv and phev with energy storage and demand response. Applied Energy. 2013;**112**:35-51

[12] Develder C, Sadeghianpourhamami N, Strobbe M, Refa N. Quantifying flexibility in ev charging as dr potential: Analysis of two real-world data sets. In: Proc. 2016 IEEE International Conference on Smart Grid Communications (SmartGridComm); Beijing, China; November 2016; pp. 600-605

[13] Althaher S, Mancarella P, Mutale J. Automated demand response from home energy management system under dynamic pricing and power and comfort constraints. IEEE Transactions on Smart Grid. 2015;**6**(4): 1874-1883

[14] Rassaei F, Soh W, Chua K. Demand response for residential electric vehicles with random usage patterns in smart grids. IEEE Transactions on Sustainable Energy. 2015;**6**(4):1367-1376

[15] Sivaneasan B, Nandha Kumar K, Tan KT, So PL. Preemptive demand response management for buildings. IEEE Transactions on Sustainable Energy. 2015;**6**(2):346-356

[16] Shafie-khah M, Heydarian-Forushani E, Osorio GJ, Gil FAS, Aghaei J, Barani M, et al. Optimal behavior of electric vehicle parking lots as demand response aggregation agents. IEEE Transactions on Smart Grid. 2016;**7**(6): 2654-2665

[17] Yu R, Zhong W, Xie S, Yuen C, Gjessing S, Zhang Y. Balancing power demand through ev mobility in vehicle-to-grid mobile energy networks. IEEE Transactions on Industrial Informatics. 2016;**12**(1):79-90

[18] Ferreira JC, Monteiro V, Afonso JL. Vehicle-to-anything application (v2anything app) for electric vehicles. IEEE Transactions on Industrial Informatics. 2014;**10**(3):1927-1937

[19] Paterakis NG, Erdinc O, Bakirtzis AG, Catalao JPS. Optimal household appliances scheduling under day-ahead pricing and load-shaping demand response strategies. IEEE Transactions on Industrial Informatics. 2015;**11**(6): 1509-1519

[20] Luo X, Zhu X, Lim EG. Load scheduling based on an advanced real-time price forecasting model. In: Proceedings of IEEE International Conference on Ubiquitous Computing and Communications (IUCC'2015); October 2015; Liverpool, UK. pp. 1252-1257

Chapter 5

Supercapacitors as Guarantors for Energy Sustainability in Low-Power Energy Harvesting Sensor Modules

Dalibor Purkovic

Abstract

Energy harvesting, low-power sensor modules are characterised by their energy independence, power consumption, size, robustness to withstand the environmental conditions, maintenance demand and long term operation. To secure any of these conditions focus has to be put on the device energy reservoir. Traditional approach would reach for the battery and at the very beginning of the development, accept the limitations that go along with it. These limitations in form of high temperature difference dependency, current peaks, limited charge cycles, loss of operating voltage and capacity, soon become constraints in the sensor module life cycle. Answer to these constraints and a guarantor of a long sensor module life cycle is a supercapacitor. An energy storage which does not have any special charging requests, other than ensuring that the maximum voltage is not exceeded, or that a minimum voltage is not reached. Supercapacitors have a low ESR (equivalent series resistance), typically of the order of 100 mΩ. This reduces internal losses during charge and discharge cycles allowing them to handle current surges without the output voltage dropping significantly. Lithium-ion supercapacitors especially have good self-discharge characteristics and retain their voltage for years.

Keywords: energy harvesting, lithium-ion supercapacitor, self-discharge characteristics, leakage current, low power, EnOcean sensor module

1. Introduction

LPWAN (low-power wide-area network) is a current research topic, alongside with the growth and popularity of the Internet of things (IoT). The main focus is the core element behind each of these aspects, i.e. the sensor node. A sensor node, which can be observed as either an independent unit, or part of a LPWAN, performs many tasks. It can harvest energy, give power to attached sensors, collect and process data from these sensors and transmit it through the air, while sustaining its own power management [1, 2]. Application fields of the IoT world are constantly expanding and with it, researched sensor platforms [3–5].

The majority of these sensor platforms rely on batteries as the main power supply. This is considered the traditional approach in sensor module energy management development [6, 7]. Problems with batteries are well known [8, 9] and thus research is now directed towards finding energy storage alternatives for use in low power sensor modules. An effective alternative is supercapacitors.

The advantages of supercapacitors are also recognised by the authors of [10]. Some of these include, very high rates of charge and discharge, little degradation of capacity over hundreds of thousands of cycles, low toxicity of materials used, high cycle efficiency (95% or more). Authors of [9] highlight the use of a supercapacitor instead of a battery to secure long term operation. This is described as the main advantage of their wireless sensor node. One of the disadvantages of supercapacitors, besides the currently higher price, is that their energy density today remains less than that of batteries, by an order of magnitude [9]. Since a supercapacitor is usually used in conjunction with a suitable solar cell, this lower energy density drawback is overcome. The potential of a lithium-ion supercapacitor is investigated in this chapter as the main energy storage for low-power, energy harvesting, wireless sensor modules.

2. Energy storage characterisation

Compared to supercapacitors, batteries have many limitations when used in outdoor sensor modules. During each transmission, current peaks in low power sensor modules can easily reach 40 mA [1, 11]. This is a serious disadvantage when utilising batteries, as they are very sensitive to current peaks. Current peaks above 30 mA limit their life cycle [8, 12], and batteries already have a lower life cycle, and are highly dependent on the number of recharge/discharge cycles. For example, the lithium-ion battery, whose characteristics are shown in **Figure 1**, can be recharged from 4000 to 7000 times. This depends on whether the previous discharge levels were less or greater than 50% respectively. After a certain period of time batteries lose their capacity and their nominal operating voltage becomes degraded, and thus will require replacement [8]. On the other hand, a

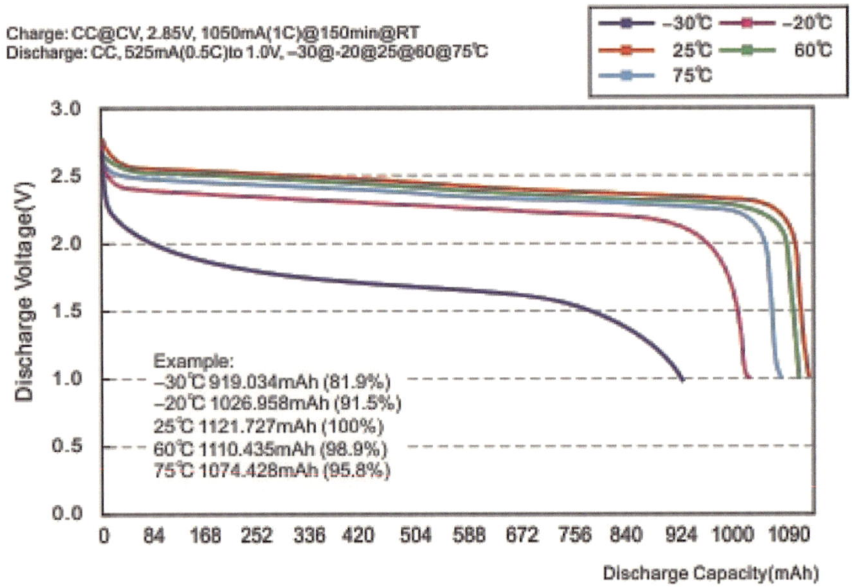

Figure 1.
Discharge temperature characteristics of the lithium-ion rechargeable battery Huahui HTC0407 with a life cycle between 4000 and 7000 times [14].

Type	Technology	Voltage (V)	Capacity (mAh)	Measured impedance (Ω)	Self-discharge	Comments
ML2430	Manganese lithium	3.1	100	≈13	2%/year	Measured 0.2 V voltage drop at 40 mA
VL2320	Manganese lithium	3.1	30	≈30	<2%/year	Too high impedance
MS920SE	Manganese lithium	3.1	11	≈35	2%/year	Too high impedance
LIR1220	Lithium-ion	3.6	8	<2	7%/month	Self-discharge values from <2%/year to 7%/month
CP1624	Lithium-ion	3.6	50	<1	20%/year	20% self-discharge per year is equivalent to steady 1 μA

Table 1.
Comparison of different technologies used for energy storage.

supercapacitor can be charged and discharged virtually an unlimited number of times. Due to this, information concerning supercapacitor life cycle is not usually provided in their specification. For example, for the used supercapacitor, the only information available regarding its life cycle, is that after 10,000 charge/discharge cycles the supercapacitor will still maintain a minimum 70% of the initially specified capacitance [13]. **Figure 1** displays a typical battery discharge curve. After full charge and start of operation there is an initial voltage drop, after which the battery voltage is stable. At a point towards the end of the lifecycle, another voltage drop will occur. The battery supply voltage will fall below the nominal voltage, and the battery will require replacement. Within the same figure, one can also see another disadvantage of batteries, the huge battery capacity temperature dependency (especially at negative temperatures).

Supercapacitors, on the other hand, do not have any special charging requests, except ensuring that the maximum voltage is not exceeded, or that the minimum operating voltage is not reached. Supercapacitors have a low ESR (equivalent series resistance); typically below 100 mΩ. This reduces internal losses during charge and discharge cycles, thus allowing them to handle current surges without the output voltage dropping significantly [8].

Different types of energy storage have been investigated. A comparison is displayed in **Table 1**. Besides internal impedance, leakage current is the most important criteria when selecting an energy store for a low power sensor module.

Based on **Table 1** and **Figure 2** it is evident that the energy storage with lithium-ion technology is the most promising for use in low power sensor modules. **Figure 2** depicts why some types of energy storage should be avoided in energy harvesting devices. The generated electrical current from a small solar cell (photo current) is shown in the same figure, assuming a worst case scenario with indoor conditions (solar cell illuminated 4 hours with only 500 lx per day). It can be seen that this harvested photo current would be completely consumed by the leakage current of, e.g., NiCad (Nickel-cadmium)-based energy storage device.

Generally, only a low impedance energy storage with minimum leakage current and sufficient capacity, is able to deliver the required energy to power-up sensors and transmit data over a longer period of time [15]. Due to limitations

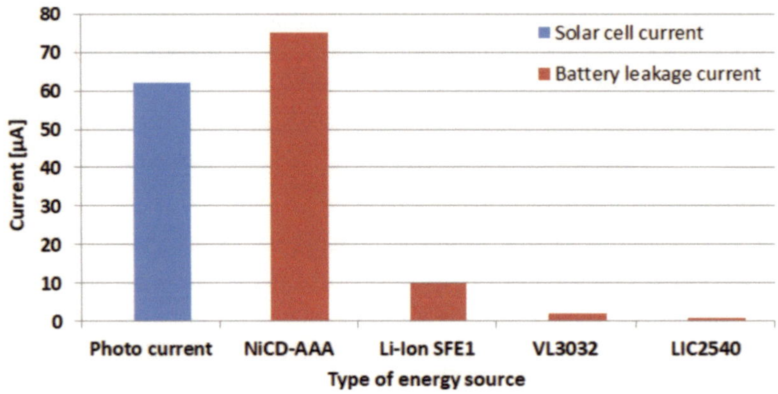

Figure 2.
Measured leakage currents of different energy stores, compared to the generated photo current from a small solar cell.

mentioned above, the tested batteries are deemed unacceptable energy storage solutions. Therefore, an alternative, with capacitor-like characteristics, is investigated.

3. Lithium-ion supercapacitors

A lithium-ion supercapacitor LIC1235RS3R8406 from Taiyo Yuden is selected as the main energy storage and became the 'heart' of the developed EnOcean low power, energy harvesting sensor module. Lithium-ion capacitors are hybrid capacitors, featuring the best characteristics of both EDLC (electrical double layer capacitors) and lithium-ion secondary batteries (LIB). Some of these characteristics are outlined in **Table 2** [13].

3.1 Self-discharge properties

Figure 3 shows the self-discharge property of the two different capacitor types. The cylinder type lithium-ion capacitor (LIC) has a 40 F capacity, when charged for 24 hours with 3.8 V, at a temperature of 25°C. The other, is a symmetrical type EDLC whose capacitance is similar to that of the lithium-ion capacitor. As seen here, the symmetrical type EDLC has a large self-discharge. After a month at 25°C, its operating voltage decreased to 80% of the initial voltage. In contrast, the LIC displays a far better self-discharge behaviour. At 25°C, it can maintain a voltage higher than 3.7 V, even after 100 days since full recharge [16]. Two additional self-discharge properties of the LIC, at two different temperatures, are also given in **Figure 4** (nominal capacity is 200 F). It is clear that after 4000 hours (at 60°C), the supercapacitor maintains close to 90% of its initial voltage. At 25°C it behaves even better; preserving 96% of the initial voltage.

The voltage retention behaviour of the lithium-ion capacitors is shown in **Figure 5**. After 22,000 hours at 25°C, this 100 F supercapacitor maintained 92% of the initial voltage [16].

The selected supercapacitor has a capacity of 40 F and this value is in the first order price compromise, compared to the 100 and 200 F versions. On the other hand a larger capacitance is not needed, since the consumption of the developed EnOcean sensor module is optimised and long term operation is secured. With the development of a more energy consuming sensor module [17], a supercapacitor with a higher capacity could be considered. The supercapacitor remains the most expensive component on the sensor module PCB, contributing to 20% of the overall costs of the assembled PCB.

3.2 Supercapacitor's leakage current

It is assumed that the sensor module due to its energy harvesting capabilities will be primarily used outdoors. Therefore, it is of great interest to define the supercapacitor's leakage current over a larger temperature range. Recorded measurements show that during a bright and sunny day, the temperature inside the sensor module housing can reach almost +70°C with direct sunlight (**Figure 6**).

	Items		Specifications
1	Operating Temperature Range (°C)		-30 ~ +85
2	Upper limit Voltage (V)		3.8
3	Lower limit Voltage(V)		2.2
4	Initial Characteristics	Capacitance	36-44F
		DCR	Under 125mΩ
5	Soldering Heat Resistance	Capacitance	Within initial spec
		DCR	Within initial spec

Table 2.
Main characteristics of the LIC1235RS3R8406 [13].

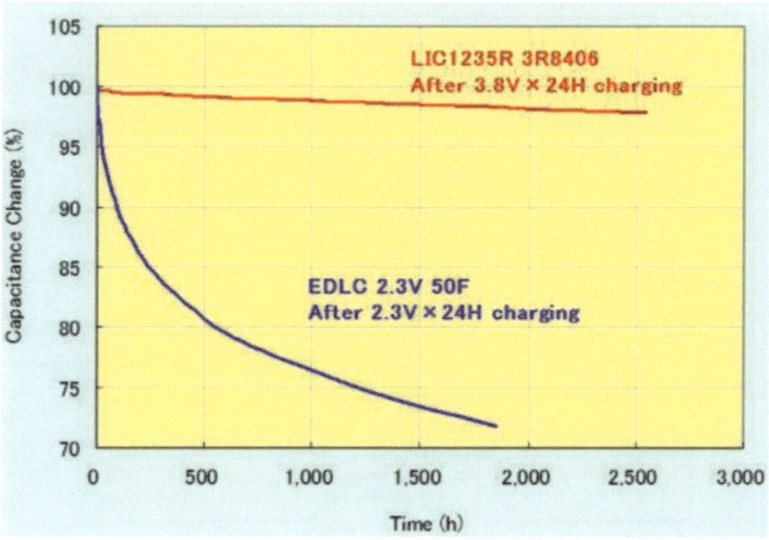

Figure 3.
Self-discharge property of the cylinder type lithium-ion capacitor with 40 F (150 mΩ) versus the EDLC with a similar capacitance [16].

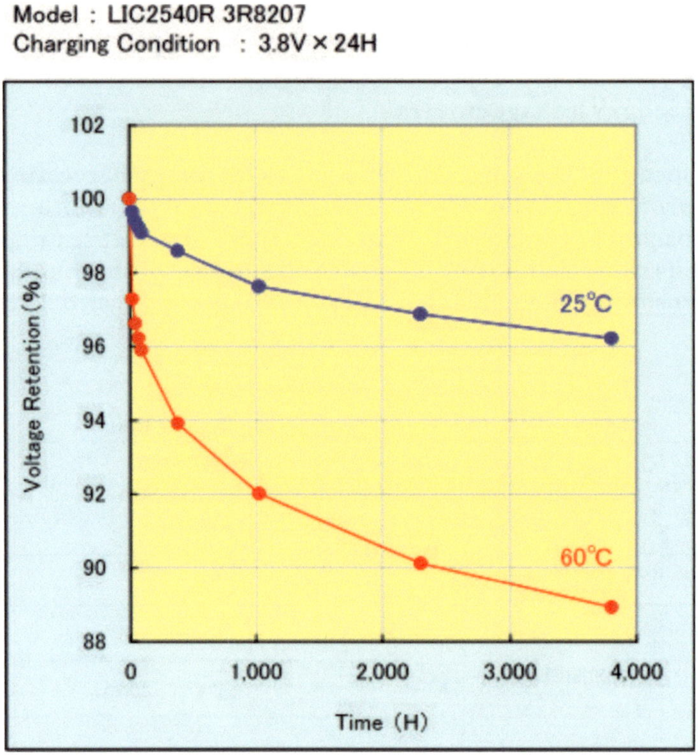

Figure 4.
Self-discharge characteristic of the 200 F (50 mΩ) Taiyo Yuden lithium-ion capacitor for two different temperatures [16].

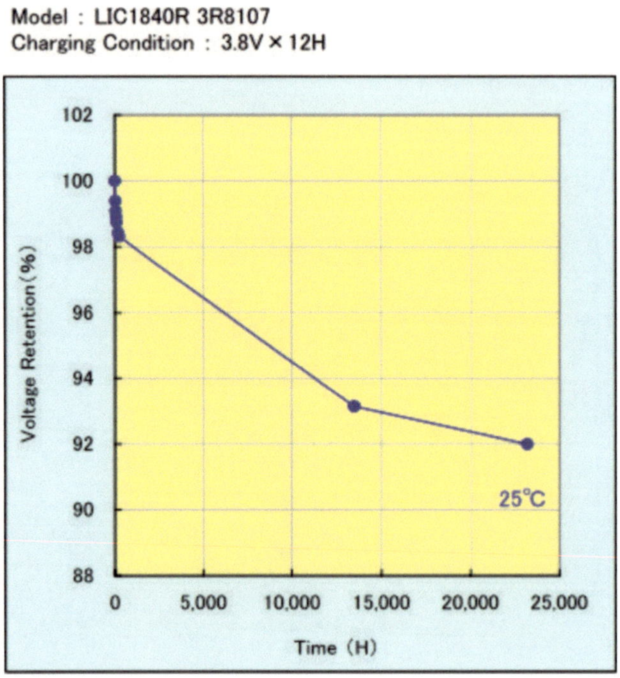

Figure 5.
Self-discharge characteristic of the 100 F (100 mΩ) Taiyo Yuden lithium-ion capacitor at 25 °C [16].

Figure 6.
Measuring temperature inside the EnOcean low power, energy harvesting sensor module housing. The two channels of the thermometer (yellow device) are measuring the temperature of the solar cell (52.3°C) and supercapacitor (50.1°C), respectively. The outside temperature was 30°C.

Temperature [°C]		85	70	60	45	25	0	-25
Leakage current after 8h [μA]	3.8 V	31.0	7.8	4.3	3.5	2.1	0.4	< 0.1
	3.5 V	12.6	3.7	1.9	0.7	0.4	< 0.1	< 0.1

Table 3.
Supercapacitor leakage current measured over different temperatures, for two different supercapacitor voltages [1].

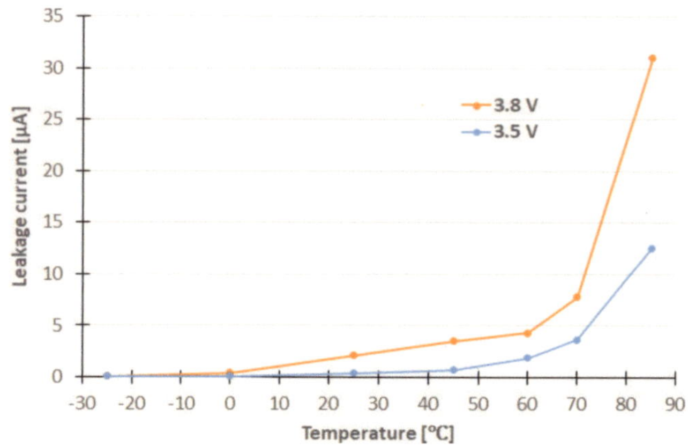

Figure 7.
*Supercapacitor leakage current values over different temperatures for different voltage levels. This is a graphical representation of the results from **Table 3**.*

A high temperature can have several negative effects on supercapacitor behaviour: the internal direct current resistance (DCR) increases, the equivalent series resistance (ESR) increases and the capacitance decreases [16]. Taking the aforementioned into consideration, the supercapacitor's leakage current over different temperatures has been measured and displayed in **Table 3** and **Figure** 7.

Measurements are taken for two different voltages of the supercapacitor; 3.8 V as the maximum usable voltage, and 3.5 V as the voltage just below maximum operating voltage. The voltage of the supercapacitor was kept constant, while the temperature was altered every 8 hours. During the last 1000 seconds of each cycle, the leakage current (current drained by the supercapacitor from the power source) was measured every 10 seconds and an average was calculated. As it can be seen in **Table 3** and **Figure** 7, a drastic increase in the supercapacitor's leakage current is observed at +85°C, compared to 25°C or lower. The supercapacitor's leakage current contributes to 40% (at 25°C) of the sensor module's overall current consumption during deep sleep mode. Therefore this must be considered when estimating dark run time operation (energy harvesting disabled).

4. The 100 Hour leakage current test

To measure the real use case behaviour of the supercapacitor and obtain its realistic leakage current figure, an additional 100 hours test was conducted. Charge and discharge cycles remaining within the supercapacitor's defined operating voltage range (from 2.2 to 3.8 V) have been executed. This was done using an automated test created with LabVIEW, as shown in **Figure 8**. For 5 hours at room temperature, a 3.8 V supply voltage was applied to the

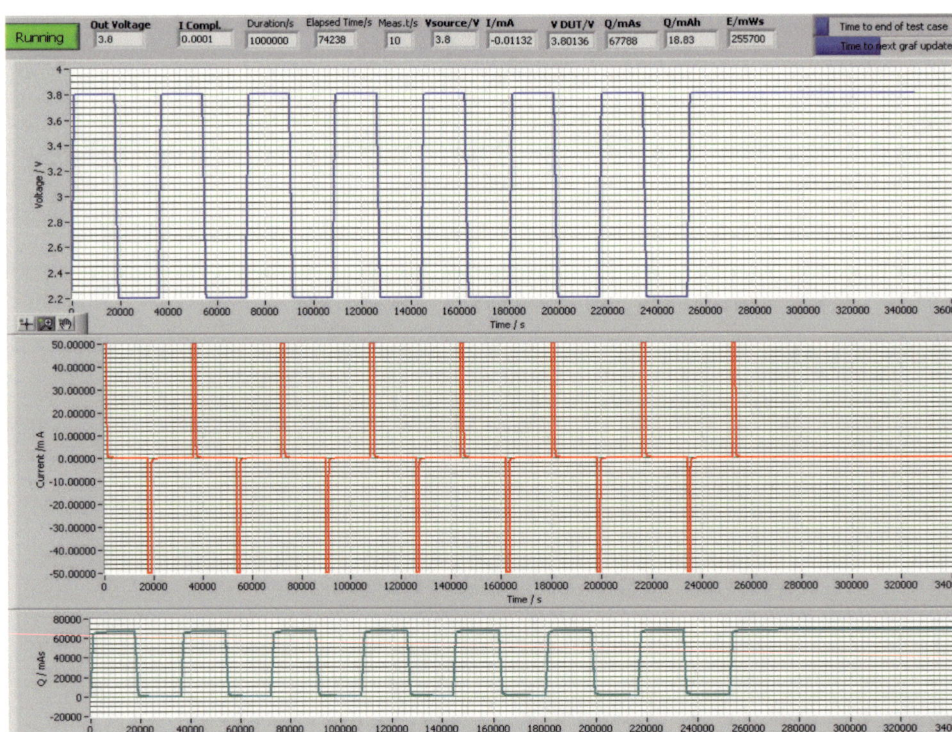

Figure 8.
Cycles of charge and discharge of supercapacitor. The blue, red and green lines represent the applied voltage, current drawn and available charge, respectively.

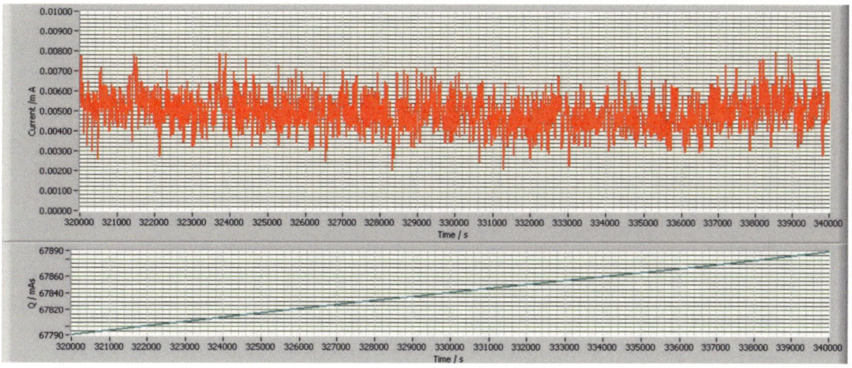

Figure 9.
Supercapacitor leakage current (red line) after 100 hours test. Fluctuations in measurement values arise due to the limited precision of the measurement equipment and noise. The green line depicts charge lost over a 5.5 hours period of time due to leakage current.

supercapacitor with a maximum current limit of 50 mA. The voltage on the supercapacitor and the charge current was recorded at 10 seconds intervals. After 5 hours, the applied voltage was changed from 3.8 to 2.2 V. The discharge of the supercapacitor then started through the power source with the capability to sink current (sink current limit set to 50 mA). This was repeated 7 times. The applied charge to the supercapacitor was also calculated, and is represented by the green line in **Figure 8**.

After seven charge/discharge cycles, a constant 3.8 V was applied to the supercapacitor for the following 30 hours and the leakage current was measured during this period.

From **Figure 9**, the average value of the supercapacitor leakage current is 5 μA. Due to this, the supercapacitor only lost 100 mC or 0.156% of its initial charge during next 5.5 hours. Couple of hours later, this value stabilised below 1 μA.

Since the recommended maximum operating voltage of most microcontrollers and radio chips is 3.6 V, the supercapacitor comes from suppliers already charged to 3.6 V ± 100 mV. It is then soldered onto the sensor module PCB and put into operation.

5. Sensor module's energy conservation

Based on the low leakage current of the supercapacitor, determined in the 100 hour test, it is expected that the sensor module will conserve its energy, for a longer period of time, when not in use. The results displayed in **Figure 10** confirm this assumption. The voltage of the supercapacitor is measured on 10 EnOcean sensor module PCBs assembled almost 3 years ago, before being measured again. These PCBs were then stored and not used. The initial voltage when these samples were first produced is represented by the red dots in **Figure 10**. The supercapacitor voltage of these 10 samples is measured again after 3 years in storage. These results are displayed by the blue dots in the same figure. It is evident that, all 10 supercapacitors lost a similar amount of charge due to their leakage current. However, the measured voltage levels are still significantly high. This suggests that 3 years after production, these sensor modules would still operate normally. In addition to this, as soon as the solar cell is connected, these supercapacitors would begin to recharge to initial voltage levels.

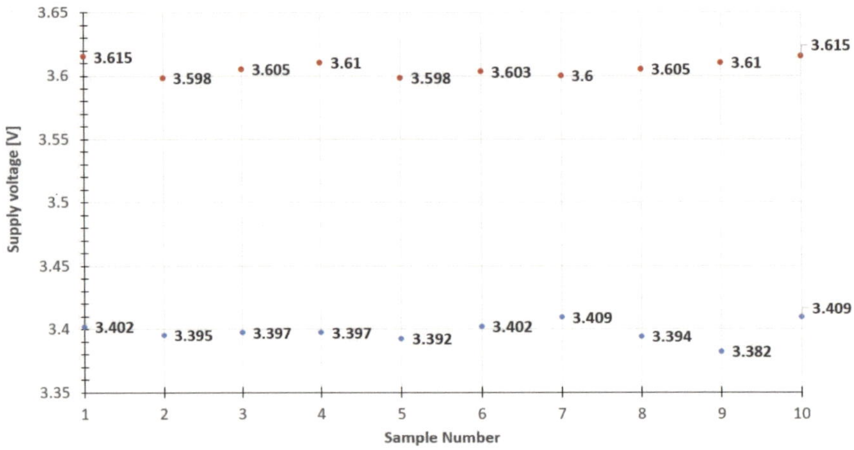

Figure 10.
Measured voltage of the supercapacitors soldered on the developed EnOcean low power, energy harvesting sensor module PCBs at initial assembly (red dots) and after 3 years of no usage (blue dots).

Author details

Dalibor Purkovic[1,2]

1 EnOcean Alliance Inc., San Ramon, CA, USA

2 EnOcean GmbH, Oberhaching, Germany

*Address all correspondence to: dalibor.purkovic@gmail.com

References

[1] Purkovic D, Hoensch M, Meyer TRMK. An energy efficient communication protocol for low power, energy harvesting sensor modules. IEEE Sensors Journal. 2018;**19**(2):701-714

[2] Keshtgary M, Deljoo A. An efficient wireless sensor network for precision agriculture. Canadian Journal on Multimedia and Wireless Networks. 2012;**3**(1):1-5

[3] Mekki K, Bajic E, Chaxel F, Meyer F. A comparative study of LPWAN technologies for large-scale IoT deployment. ICT Express. 2019;**5**(1):1-7

[4] Wang N, Zhang N, Whang M. Wireless sensors in agriculture and food industry—Recent development and future perspective. Computers and Electronics in Agriculture. 2006;**50**:1-14

[5] Guibene W, Nowack J, Chalikias N, Fitzgibbon K, Kelly M, Prendergast D. Evaluation of LPWAN technologies for smart cities: River monitoring use-case. In: IEEE Wireless Communications and Networking Conference (WCNC); San Francisco: CA; 2017

[6] Mendez GR, Yunus MAM, Mukhopadhyay SC. A WiFi based smart wireless sensor network for monitoring an agricultural environment. In: IEEE International Instrumentation and Measurement Technology Conference (I2MTC); Graz: Austria; 2012

[7] Vellidis G, Tucker M, Perry C, Kvien C, Bednarz C. A real-time wireless smart sensor array for scheduling irrigation. Computers and Electronics in Agriculture. 2008;**6**(1):44-50

[8] Beeby S, White N. Energy Harvesting for Autonomous Systems. Norwood, MA, USA: Artech House; 2010

[9] Simjee F, Chou PH. Everlast: Long-life, supercapacitor-operated wireless sensor node. In: ISLPED; Tegernsee: Germany; 2006

[10] Voorden AMV, Elizondo LMR, Paap GC, Verboomen J, Sluis LVD. The application of super capacitors to relieve battery-storage systems in autonomous renewable energy systems. In: IEEE Lausanne Power Tech: Lausanne; 2007

[11] Morin É, Maman M, Guizzetti R, Duda A. Comparison of the device lifetime in wireless networks for the internet of things. IEEE Access. 2017;**5**:7097-7114

[12] Bardyn JP, Melly T, Seller O, Sornin N. IoT: The era of LPWAN is starting now. In: 42nd European Solid-State Circuits Conference (ESSCIRC); Lausanne: Switzerland; 2016

[13] Taiyo Yuden cylinder type lithium ion capacitors [Online]. October 2018. Available from: https://ds.yuden.co.jp/TYCOMPAS/or/download?pn=LIC1235RS3R8406&fileType=CS [Accessed: February 2019]

[14] Huahui Energy. Super Li-ion battery [Online]. December 2017. Available from: https://www.amec-gmbh.de/wp-content/uploads/2017/12/Huahui-Energy-Super-LI-Ion-Battery-2.pdf [Accessed: March 2019]

[15] Martinez B, Montón M, Vilajosana I, Prades JD. The power of models: Modeling power consumption for IoT devices. IEEE Sensors Journal. 2015;**15**(10):5777-5789

[16] Taiyo Yuden. Taiyo Yuden lithium ion capacitors: An effective EDLC replacement [Online]. Available from: https://www.yuden.co.jp/include/english/solutions/lic/LIC_White_Paper_Final.pdf [Accessed: February 2019]

[17] Purkovic D, Coates L, Hönsch M, Lumbeck D, Schmidt F. Smart river

monitoring and early flood detection system in Japan developed with the EnOcean long range sensor technology. In: SMAGRIMET 2019—Second International Colloquium on Smart Grid Metrology. Croatia: Split; 2019

Chapter 6

Energy Management through Electromagnetic Conversion

Eduardo Torres-Sánchez

Abstract

The global society has the responsibility to concern about environmental impact for energy purposes by replacing existing coal and hydrocarbon methods by sustainable and efficient energy systems. Hence, current power generation systems are bounded by the physical laws that tend to decrease the performance by converting most of the energy into heat. Likewise, the revolution and massive implementation of renewable energies around the world have demonstrated that the electromagnetic transduction presents a viable option to harness the induced mechanical energy provided by either wind or water into exergy. The exergy focuses on the efficiency of the second law of thermodynamics with the purpose to ensure availability and quality of energy within energetic management systems. Thereby, it is necessary to decrease the energy demand by making very efficient power-consuming devices and increasing the quality of energy with performed output power generation systems. This chapter addresses a single diagram to develop models and novel designs for power generation with the aim to develop variable efficiency power systems. Furthermore, an analysis is addressed on magnetism, electromagnetic induction, and magnetic materials to design, optimize, and implement into current power cycles.

Keywords: variable energy efficiency, electromagnetism, sustainability, energy quality, time-varying magnetic field

1. Introduction

Energy management mainly refers to developing systems with potential energy savings with a positive economic impact. The key relies on identifying energy saving opportunities [1], for instance, the accurate maintenance measures made on time and the necessary modifications of current power systems to ensure best operation at lower costs. Accordingly, renewable energies have opened an opportunity to increase power management efficiency without compromising the natural resources of future generations [2].

Nevertheless, the energy is classified by two variants stated by the quality of energy, named exergy and anergy. The term exergy was first mentioned by Zoran Rant in 1956 as the amount of energy that can be converted either into mechanical, electrical, or other work [3]. Thus, the anergy results as the remaining energy part without a practical utility. Moreover, the exergy is a measure of efficiency of an energy system into parameters of quality and availability [4]. In comparison, these statements are based on the second law of thermodynamics that establishes the

 IntechOpen

boundaries regarding the quantity of work that can be done. In addition, an exergy analysis compares different energetic performance to choose the most efficient alternative according to either storing, consuming or power generation application [5].

Likewise, the exergy determines the thermodynamic value in a quantitative manner by analyzing the energetic resource wasting to find out the causes of low efficiency. Thereby, once the causes are quantified, the exergy analysis can help to specify the necessary modifications on either the process or the design [6]. Thus, the exergy is stated as a thermodynamic property of a substance or a system that allows determining the useful potential work of an available amount of energy that can be acquired by the spontaneous interaction between a system and the environment that surrounds it [7].

On the other hand, the only way to operate larger systems is across electrical energy provided by electric power plants. Hence, electricity generation has been accomplished in the last century by nuclear methods, coal and fossil-fueled systems, largely designed to power supply to an endless number of power consumption devices [8]. However, regardless of these great energy structures that produce huge amounts of useful energy, they face environmental concerns by polluting the natural resources [9]. In addition, it is not only the pollution but the consequences of extracting oil more than what it is due, by triggering greater earthquake frequency and intensity [10]. The oil is the earth's lubricant. Moreover, the environmental impact of using the hydraulic fracking method to extract shale gas causes an irreversible damage to aquifers and subsoil [11]. However, an additional problem despite the power generation from hydrocarbons is the need of power supply systems to be independent from the grid.

In contrast, renewable energies have started to accomplish the energetic demands with the aim to stop polluting by converting the mechanical energy from natural sources such as wind, wave, geothermal, hydraulic, and tidal power into available electrical energy [12]. The energy conversion is made, thanks to the electromagnetic induction properties that are present in every time-varying magnetic field of current electric generators. Of course, while there is an existing mechanical energy, consequently, there will be energy for conversion. Indeed, if there is no energy to convert, then no energy is generated. This statement is underpinned in the first law of thermodynamics [13].

Otherwise, energy harvesting systems have become an alternative form of power generation over the last few years. The mechanical force induced by vibrations using piezoelectric materials, and even collisions, are examples of energy harvesting methods. However, electromagnetic induction is an especially promising means of energy harvesting, since only coils and magnets are needed for its functioning and efficiency increasing [13]. Thus, the electromagnetic transduction increases the performance of energy conversion more than hydrocarbons by the simple fact of combustion and excessive heat conversion energy lack. This change yields implicitly the performance of the second law of thermodynamics, which is the aim of the exergy, by developing more efficient power systems [14]. Another important fact to regard is that the increase in the energy demand is on the rise since the population is on the rise [15].

Accordingly, it is necessary to power supply medium and low power consuming systems by obtaining the energy from the environment instead of getting it from a big manager energy such as oil wells. Thus, the development of new energy systems technology must be every time more efficient to power supply without environmental impacts at the lowest cost.

In this chapter, a developing diagram has been addressed to design customized electric machines according to the desired applications. The applications are also focused to establish these generators in areas where it is difficult to supply energy for common use electronic devices and in applications of everyday life. Of course, this effect would yield to create independent power grids per zones, roads, and streets. For instance, the sum of renewable energy applications will determine the approach to the creation of new energy systems.

Henceforth, these designs will open the possibility of developing configurable generators, according to the real-life applicable power generation systems. The effectiveness of the design depends on the architecture of the device, while the quantity of energy harvested depends on the lifetime of the magnets and on the continuation of the induced mechanical force.

Elsewhere, as the exergy focuses on the availability and the efficient application of energy, this statement yields to adapt current power generators into more efficient power cycles with the aim to reduce the energy transformation into heat. For instance, current induction generators of wind power stations may have the enhancement to vary the distance between the rotor and the stator with the objective to harvest the energy of wind velocities beyond the current operating limit. Therefore, the rotor would vary the distance from the stator every time there exists an up-or-down variation of the wind velocity. Therefore, at low wind velocity, the rotor will be set up farer from the stator to generate low output power and vice versa; at greater wind velocities, the rotor will be set up at a closer distance from the stator to generate high output power. Moreover, this variability allows velocity fluctuations to harvest energy any time there is an induced mechanical force. Namely, the outcome is a variable efficiency generator in accordance with the mechanical energy provided by the environment.

Further, there exist many energy systems that can be modified to increase efficiency by reducing the energetic losses with the induction of electromagnetism in a performed manner.

2. Energy management in energy harvesting

Renewable energies are an essential support to provide exergy in a clean manner to human life. Nowadays, an increasing number of smaller technologies are being powered by batteries, renewable energies, and complex control systems to save energy. Electromechanical applications are currently focused on energy harvesting, including the development of autonomous devices [16]. However, the current energy harvesting methods still rely on batteries or their equivalent. Namely, neither fossil fuels nor nuclear energy fall into this category. Thus, energy harvesting can play an important role if used properly in power systems.

Building on single-source systems, Amanor-Boadu et al. [17] created a multi-energy system which simultaneously charged a Li-ion battery. This idea, however, could now be further improved by adding more energy sources to one energy harvesting system and not only charge one Li-ion battery or one capacitor but rather several batteries in less time.

A diagram of a multi-energy system is shown in **Figure 1** as an effective solution to combine multiple energy harvesting systems simultaneously for battery charging [17]. Evenly, this approach should have an effective management energy system such as a BMS (Battery Management System) to protect the batteries inside their safe operating functionality. Moreover, the effectiveness decreases, because energy is lost during the conversion processes. However, electromagnetic induction could increase the efficiency [18].

On the other hand, the exergy presents different qualities that depend on the possibility either to generate work or transforming one sort of energy into others. For instance, the heat quality depends on the temperature, where at greater

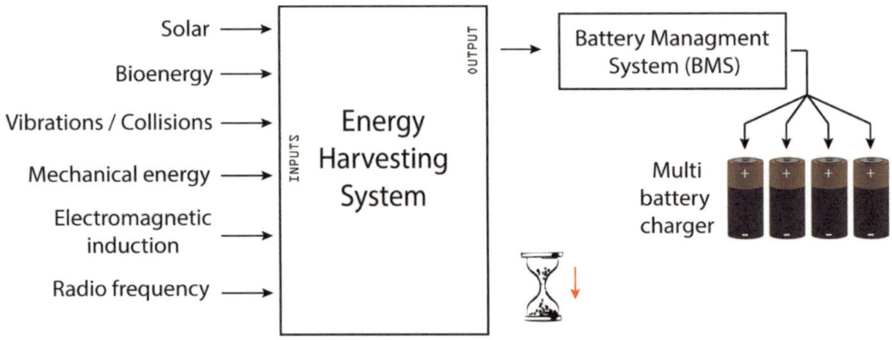

Figure 1.
Multi-battery charging system with multiple energy sources.

temperature, a heat source can transfer its energy more easily than at lower temperature [19]. Commonly, it is accepted as a measure of energy quality, the capacity of an object to produce work. Thus, for a thermal machine to perform work, the heat must be taken from a power source at high temperature, and part of that heat must be transferred into a low temperature environment while the thermal equilibrium is being carried out. In comparison with a thermal machine, if the environment temperature is very low (cold), therefore, it results more difficult to transform the heat of this source into work. Therefore, the reference level (the value of the low temperature) is very important when defining the exergy. This physical phenomenon is known as entropy, describing the irreversible for a thermodynamic system in equilibrium [20]. Consequently, as thermal machines usually work with the surrounding medium as a cold focus, the reference level is then taken from either a room or environmental temperature [21].

Accordingly, because of the lack of thermal equilibrium in the environment, the reference state cannot be completely defined but it is enough by defining the thermal equilibrium through temperature. The exergy of a substance can be divided into four main components: kinetic, potential, physical, and chemical exergy. The last exergies, physical and chemical, are grouped by the thermal exergy which is the sum of both. Conversely, the effect of energy losses during the energy conversion from mechanical to electrical through electromagnetic transduction is much more lower than in thermal machines [7].

Finally, researchers often compare the effectiveness of different methods based on the energy storage density inherent to each transducer type, demonstrating that electromagnetic induction demonstrates better performance than electrostatic [14]. The most effective transducer type depends on the specific structure design, the implemented materials, and its application.

3. Electromagnetic induction

In 1831, Michael Faraday and Joseph Henry discovered ways to produce electricity from magnetism—one, by using one long coil called an intensity magnet and the other by passing a magnet inside a short coil called a quantity magnet. These discoveries became the most important research on electric and magnet induction [22].

The multi-atomic arrangement of magnetic structures of the individual magnetic momentums of a group of atoms/molecules stays aligned due to a strong coupling named domains or magnetic dipoles [23]. The electron motion of the atoms has several domains. Further, the intrinsic magnetic dipole moment is associated with the spin of the electrons. Thus, the alignment of magnetic dipoles parallel to an external magnetic field increases the field.

Equally, the difference by comparing the magnetic field lines of a magnetic dipole with the electric field lines of an electric dipole is the direction [24]. In other words, inside the current loop, the magnetic field lines are parallel to the magnetic dipole moment, whereas among the charges of the electric dipole, the electric field lines are opposite to the direction of the dipole moment. Thereby, inside a magnetically polarized material, the magnetic dipoles create a parallel magnetic field to the magnetic dipole moment vectors [25]. Nonetheless, if the magnetic flux across the stationary wire loop is changing, an electromotive force (emf) is induced in the loop.

The emf is distributed throughout the loop, which is due to nonconservative electric field tangent to the wire as shown in **Figure 2**. The flux across the loop is changing because the magnetic field strength is increasing, so an emf is induced in the loop. Since emf is the work done per unit change, we know that there must be forces exerted on the mobile charges doing work on them [25].

Otherwise, Lenz's law does not specify just what kind of changes cause the induced emf and current. The induced emf is in such a direction as to oppose or tend to oppose the change that produces it [26].

From **Figure 2**, when the bar magnet is moving to the right, an induced current is generated since an emf is induced into the loop. Likewise, the magnetic field through the induced current into the loop produces a magnetic field that exerts a force on the bar magnet by opposing its motion to the right. For instance, the yellow arrow is taken as if it were a bar magnet, where the magnetic moment of the loop $\vec{\mu}$ is such as to oppose the motion of the bar magnet due to the induced current. Indeed, the bar magnet is moving toward the loop, so the induced magnetic moment repels the bar magnet [25].

3.1 Permeability, reluctance, and magnetic susceptibility

When a magnetized material is placed in a strong magnetic field, such as a coil or solenoid, the magnetic field of the coil tends to align the magnetic dipole moments inside the core [25]. The magnetization occurs, thanks to the microscopic current loops inside the magnetized material. These current loops are a classical model for the orbital motion and spin of the electrons in atoms.

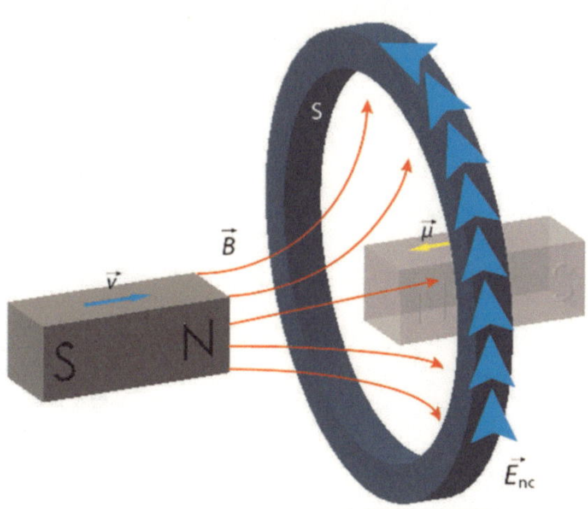

Figure 2.
Single stationary loop in a magnetic field.

The magnetic susceptibility X_m is a dimensionless proportionality constant that defines the susceptibility degree to the magnetization of a material influenced by a magnetic field. This term is related with the permeability of the materials. Thus, the magnetization of ferromagnetic materials exhibits magnetization even in the absence of an applied field. The magnetic susceptibility of the copper, implemented to make the coils, is $-0.98 \times 10^{-5}\ X_m$ at 1 atm. Moreover, silver and gold would be a better option as conductors to harvest energy with electromagnetic variations due to the values of their magnetic susceptibility which are -2.64 and $-3.5 \times 10^{-5}\ X_m$, respectively [24]. Additionally, the magnetic permeability μ is defined as the ability of a material either to attract or pass across magnetic fields [27].

Consequently, there exists an interaction between the density of the magnetic field and the magnetic induction that appears within itself. Likewise, the magnetic permeability of the medium can be defined as the capacity measure to establish lines of magnetic flux. Further, the greater the permeability of the medium, the greater the number of flow lines per unit area as shown in **Figure 3**.

Elsewhere, the magnetic permeability of the air or vacuum is $4\pi \times 10^{-7}$ Wb/Am, Tm/A or H/m, represented as μ_0. Thereby, regarding the air permeability as a reference, the relative permeability μ_r of any material will be measured respect to it. For instance, copper has 0.9, iron is between 1500 and 7200, and NdFeB is over 100,000 H/m [27]. Thus, permeable materials are magnetized by magnetic induction, resulting in a much more intense magnetic field, and accordingly, the use of these materials increase the efficiency of power generation.

In contrast, the reluctance is the opposite of the permeability, in which, this applies resistance to the magnetic flux when influenced by an external magnetic field. The greater the reluctance of a material, the more energy will be required to establish a magnetic flux through it. Thereby, these properties allow an efficient conversation of energy by increasing the quality and availability of energy.

3.2 Magnetic materials

The magnetic flux produces electric currents and vice versa. The fields generated by magnetic materials are because of the orbital angular momentums and the electron spinning. The continuous movement in the material experiment forces ahead an applied magnetic field. The magnetic characteristics can vary by the composition of other elements, where the atomic interactions are modified [24].

The applied magnetic field always plays an important role over the regarded electrons individually, giving the effect known as diamagnetism. At an atomic level, the magnetic momentum is aligned with the induced field, giving place to

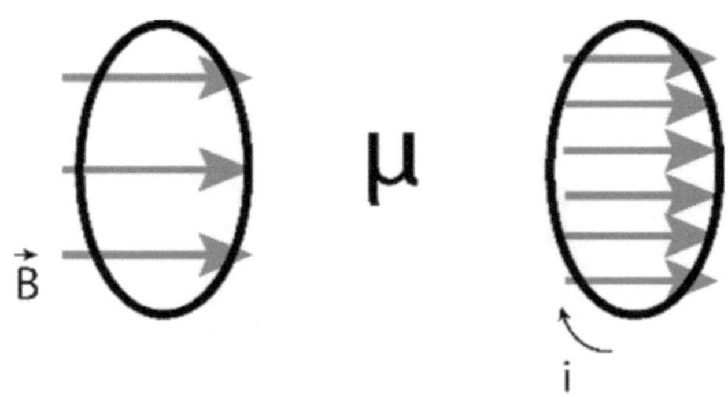

Figure 3.
Permeability behavior to magnetize the materials [27].

paramagnetism. Simultaneously, the present thermic energy orients in a random way the magnetic momentums. This interaction, as a relative intensity of all these effects, determines the definitive behavior of the materials [28]. According to the behavior of their magnetic moments in an external magnetic field, the materials fall into three categories, which are paramagnetic, diamagnetic, and ferromagnetic [25].

The diamagnetic materials have the property to induce very small magnetic moments compared with the permanent magnetic moments. This effect arises from the orbital magnetic dipole moments induced by the applied magnetic field. The magnetic moments are opposite the direction of the applied magnetic field [29]. The diamagnetism effect can be modeled by applying Lenz's law to the orbital movement of the electrons. Some examples for diamagnetic materials are copper and helium.

In comparison, paramagnetic materials arise from the molecular magnetic moments by an applied magnetic field. Equally, the magnetic dipoles interact with each other and they are oriented in a randomly way. The magnetic momentum has a parallel alignment to the applied field. Since the slow response, these materials are similar to the air ($\mu = \mu 0$) in the magnetic design. Moreover, the response intensity is very small, and the effects are practically impossible to detect, except at extremely low temperatures or very intense applied fields. For instance, aluminum and sodium present these characteristics [30].

Further, ferromagnetism is more complicated. The strong interaction among external magnetic fields and the ferromagnetic material causes a very large increase in the field. A high degree of alignment occurs even with weak external magnetic fields. The magnetic momentum of big atom groups remains aligned with each other due to a strong coupling.

A disadvantage with these materials is the high temperatures, in which it tends to misalign the domains. This temperature is named Curie's temperature T_c (K), becoming a paramagnetic material due to the disordered thermic effects greater than the alignment effects of the magnetic interaction among the domains. Therefore, to demagnetize a magnetized material, it is only necessary to heat up over the T_c [31]. For instance, the *Fe* can be demagnetized beyond 1043 T_c (K). Other examples of ferromagnetic materials are cobalt, iron, nickel, and most of the steels.

Hence, the permanent magnets have their magnetic poles aligned generating an external magnetic field. Many permanent magnets are made by metallurgic techniques, where the material is milled until it is converted into small dust particles. The magnets not only generate either an own or induced magnetic field but continue producing an induced magnetic field even after the applied magnetic field is retired. This property is neither altered nor weakened with time, except when the magnet is subjected to high temperature changes, demagnetized fields, and mechanical tensions, among other situations [24]. For instance, **Figure 4** shows the alignment of the magnetic field of a ring-shape permanent magnet with diamagnetic materials such as copper.

Elsewhere, only in the transition elements, such as Fe, Ni, O, and the rare earths elements, there exist incomplete deep orbitals that are not affected by the bonding forces when the atoms come together to form a solid. The atoms retain an important magnetic momentum and this effect yields to the origin of the phenomenon of ordering [32].

In contrast, the electromagnetic induction increases the quality and availability of energy harvesting by including permanent magnets and coils. Roughly speaking, magnetic materials are implemented because they have very large positive values of magnetic susceptibility. The reason is because in a changing magnetic field and in a changing magnetic flux, an emf is induced [22]. This feature helps harvesting the energy of the magnet in a singular manner by changing the magnetic flux in several

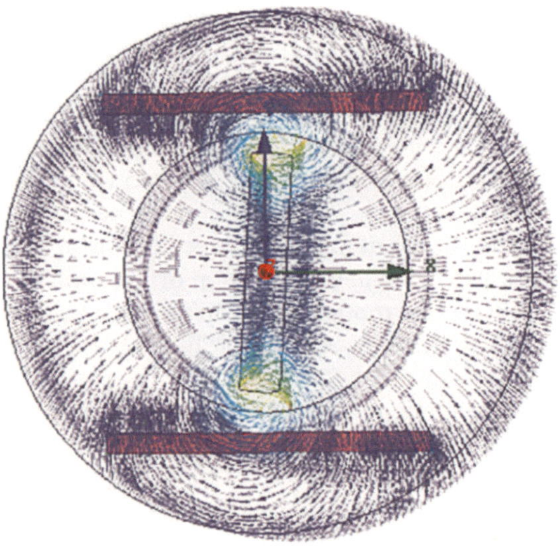

Figure 4.
Magnetic field distribution between a permanent magnet (center) and a diamagnetic material (top and bottom).

ways. Thus, the quantity and availability of exergy increase as a viable option to entirely substitute fossil fuel energies by renewable energies through the electromagnetic conversion.

3.2.1 Ceramic ferrites and neodymium magnets

Ferromagnetic materials have been regarded as highly important electronic materials for more than half a century. During this time, the characteristics of commercial ferrite materials, both soft and hard ferrites, have come to approach theoretical values. The quality of commercial ferrites has been improved through accumulated scientific knowledge and advanced technology [33]. Further, when the ferrites reach the boundary magnetic saturation B_{sat} (**Figure 6**), is because the magnetic momentums of all the particles are entirely aligned. Consequently, so that it happens, it is necessary to induce much more energy to the material by increasing the cost. However, the advantage to magnetically saturate the whole material is low, since there exists a small difference by not to doing so.

The ceramic ferrites are made by using iron oxide powder. The formula is X_n (Fe_2O_3), where X can be either B or Sr. and $5.8 < n < 6.0$ to improve the alignment of the crystal structure. After the milling, the powder is compressed in a matrix with a magnetic field applied. The compacted powder is then synthetized at 1100–1300°C [31]. Further, the permanent magnetism in ceramic ferrites is based on the anisotropy magnetocrystalline. **Figure 5** shows the demagnetization diagram of these materials.

On the other hand, the production of neodymium-iron-boron is cheaper than producing cobalt-samarium magnets. Indeed, iron is a transition metal much cheaper than cobalt, and neodymium is a rare light earth which is much more abundant than samarium. In various tests, boron formed a ternary compound with strong uniaxial magnetocrystalline anisotropy, demonstrating a higher operating temperature [34].

Additionally, the approximated formula is $Nd_2Fe_{14}B$, which states the best combination of magnetic and thermic properties. These magnets have different combinations in Nd and Fe proportions, producing a wide range of available properties.

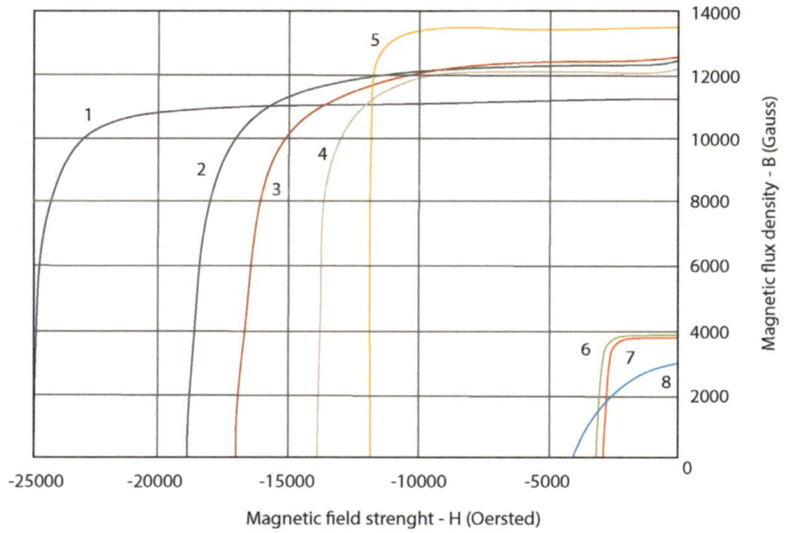

Figure 5.
Demagnetization graph of ceramic ferrites and NdFeB magnets [35].

These properties are presented in different values of magnetic flux (G) and magnetic field strength (Oe); **Figure 5** shows the demagnetization curves for various combinations of NdFeB [35].

3.2.2 Demagnetization of magnetic materials

The demagnetization of ceramic ferrites and NdFeB magnets is compared, where **Figure 5** shows the powerful magnetic flux density and magnetic field strength of NdFeB magnets against the ceramic ferrite magnets [36].

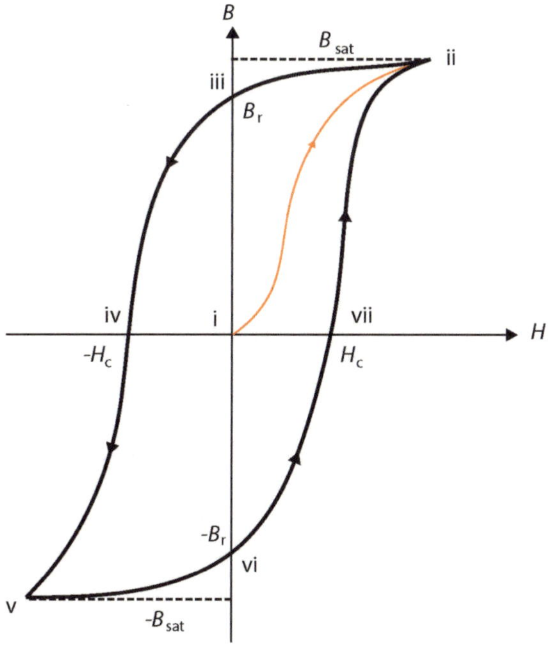

Figure 6.
Initial magnetization curve with magnetic saturation state [32].

According to the demagnetization graph, the materials represented by the numeric value are: (1) $Nd_{31}Fe_{25}B$, (2) $Nd_{35}Fe_{19}B$, (3) $Nd_{38}Fe_{17}B$, (4) $Nd_{40}Fe_{14}B$, (5) $Nd_{44}Fe_{12}B$, (6) $SrFe_{12}O_{19}$, (7) mild steel, and (8) molten iron. Meanwhile, to change the Gauss units to Tesla, it is only necessary to multiply it by 10^{-4} T. Thereby, it is necessary to regard that the maximum magnetic field B in a ferrite magnet $SrFe_{12}O_{19}$ is 3900 G with a magnetic field strength of 3450 Oe [37].

The normal magnetization curve determines the magnetic flux density B that the magnet generates according to the demagnetization force value H. As H is closest to the reversion value $-H_c$ (**Figure 6**), the flux density decreases until it drops quickly by passing the crest of the curve. Thus, in practical applications, the magnet should be kept over the crest to obtain a useful magnetic flux [32].

Nevertheless, when selecting a material for permanent magnet uses, the idea is to have three characteristics such as high remanence to obtain a greater magnetic flux, high coercivity to avoid the magnet from demagnetizing easily, and the maximum energy point to require less material to generate the magnetic flux [32]. Of course, it is not possible to possess these characteristics simultaneously. Likewise, a hard material with high remanence and coercivity presents a hysteresis cycle of great surface, implying high anergy. Therefore, this is the reason to use soft materials that use alternating current, resulting in narrow hysteresis cycles and lessen losses. Further, the hard materials are utilized in applications where the permanent magnets are not exposed to magnetizing-demagnetizing cycles.

4. Magnetic hysteresis

Demagnetized materials can be magnetized when the material is placed in an existing magnetic field space. Indeed, the magnetization varies when the applied field varies. Moreover, a magnetic domain rotation of the material starts as the magnetization process occurs. Thus, the rotation domains yield to align with the applied field. This process has lessened energy consumption; hence, the magnetization curve has a rapid increment of magnetic values. Additionally, the next step is the orientation of the magnetic domains which have not been completely aligned yet. The process involves a greater expenditure of energy, and thereby, the magnetization curve increases slower. Thus, it comes to a moment where all the domains of the material are aligned with the applied field and this final process is named magnetic saturation [24].

The nonlinearity magnetization curve consequently yields to magnetic domain deformation due to the thermodynamic characteristics as well as the interaction between each other [38].

According to **Figure 6**, the demagnetized state (i) indicates that every time the emf H arises, more domains are being parallelly aligned until all of them are aligned in the saturation state (ii) where there exists an induction field B_{sat}. At this point, if the emf is increased, then no more alignments will occur. In contrast, if the induced emf is decreased due to the saturation state (ii), the system does not follow the same trajectory. The reason is because the alignment domain mechanism, the border domain movements, and the thermal agitation are highly nonlinear mechanisms.

Equally, when the emf is equal to zero (iii), the material stays magnetized, generating a residual induction field B_r known as remanence. Hence, if the emf increases with negative values, the material is demagnetized effectively until reaching the coercivity value $-H_c$ (iv). Thereby, a new saturation is generated but in the opposite sense (v). This behavior is repeated over a symmetric curve in (vi) and (vii) sections [32].

The energy systems follow the same pattern, known as the hysteresis cycle. The material magnetization is made at the expense of energy, dissipated in heat form due to the border domain alterations. Furthermore, when a hysteresis cycle takes place in a material, it experiments an energy delivery by volume unit in heat form, equal to the hysteresis cycle [24]. However, the energy losses are much lower than in a thermal machine.

Additionally, the ferrite magnet keeps a magnetization $+B_{sat}$ until an inverse field of magnitude $-H_c$. Thus, the magnetization becomes unstable and decreases to $-B_{sat}$. Thereby, a new field $+H_c$ is required to apply so that the magnetization increases to $+B_{sat}$. Likewise, the first quadrant represents the initial magnetization region, whereas the second quadrant represents the region in which the magnet does the work against an applied reverse field with a lower value than $-H_c$ [32].

The presented plane in **Figure 6** has three main considerations for the technological application design [32]:

- Maximum energy point BH_{max}, which is exactly at half way between B_r and $-H_c$ over the second quadrant. This property has a value of $BH_{\mathrm{max}} = \mu_0\ (\frac{1}{2}M_{\mathrm{sat}})^2$. It also represents the maximum energy density that a magnet can stock.

- Coercivity H_{c} is the intersection of the curve with the $-H$ axis. An ideal material would be $H_{\mathrm{c}} = M_{\mathrm{sat}}$; however, an emf is required to set aside the magnetic flux inside the magnet. This property states the capacity of a magnet to stand demagnetizing factors.

- Remanence B_r is the intersection with $+B$ axis. An ideal material has $B_r = \mu_0 M_{sat}$. Nonetheless, B_r is the magnetic flux density value when the magnet has not emf ($B_r \rightarrow H = 0$). The remanence is an index of the capability of the material as permanent magnet.

Henceforth, the behavior of a magnet can be described and restricted to the second quadrant of the demagnetization curve.

5. Development diagram for electromagnetic systems

Ferromagnetic materials have a relative permeability of several hundred thousands over paramagnetic and diamagnetic materials. Indeed, these materials are strongly attracted by external magnetic fields. For instance, the permanent magnets and the iron structure (stator) in which the coils are placed inside an electric generator cause the effect to generate greater and denser time-varying magnetic fields according to the magnetic flux and electric fields characterized by the magnetic hysteresis and attributed to the magnetic alignment dipoles.

Namely, the main steps to design electromagnetic energy systems are shown in a cycling way in **Figure 7**. It is necessary to use all the physical laws and elements written inside the circle so that any design works properly. Usually, this scheme may regard either rotational or linear oscillatory velocities.

Specifically, each step is described as follows:

a. Select the magnet type and magnet shape to use in the core. This is the most important step because from here, the rotation direction (if any), coil position, and basically the whole design are defined. Although the three magnetic characteristics are not possible to be present simultaneously, either select or if possible design a permanent magnet with the closest characteristics of high

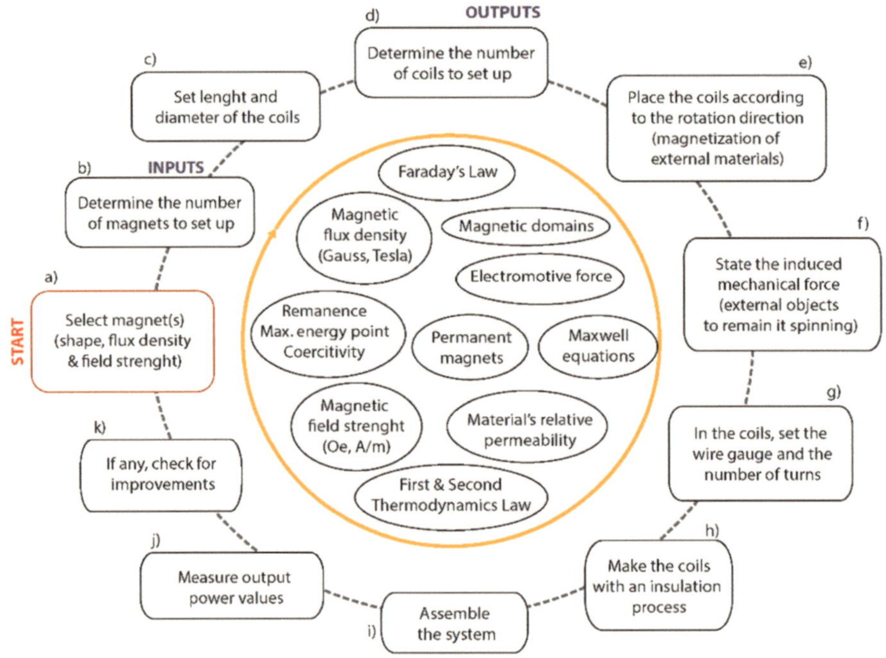

Figure 7.
Diagram for power generator design.

coercivity (to avoid the magnet from demagnetizing easily), maximum energy point (to require less magnetic material to generate a magnetic flux), and high remanence (to obtain a greater magnetic flux).

b. Determine the number of magnets (inputs) and poles. The selection validation is related with both, the desired values and design to satisfy the energetical necessities. If it is designed for low power generation, then, a small magnetic value is suggested and vice versa. At this point, the electromagnetic machine may be designed as a variable efficiency energy system to harness most of the induced mechanical energy as much as possible, for instance, the variable distance among the magnets and coils.

c. Set the preferred diameter as well as the desired length of the coil. These measures are proportional to the magnet size. If in the simulation, the system has a low quantity of magnetic dipole alignment, then it is suggested to perform the design. Nonetheless, if it is designed for high power applications, then high magnetic values are desired as well as a greater mechanical energy. Of course, the design can be set by shunts instead of coils. This kind of design will deliver a very efficient energy system and therefore, good quality and availability of energy.

d. Determine the total number of coils (outputs). This step will determine the number of preferred phases and the consecutive connection in which the energy system will be generating energy. In this point, an exergy analysis can be made to study the energy that will be lost during the hysteresis of the system. Both thermal conditions and turns of the induced coils must be regarded as well.

e. Place the coils according to the rotation of the magnet. The coils can be placed either horizontally or vertically. The aim is to get the best position for the straightest magnetic alignment of the materials in each rotation.

f. State the mechanical force induced to the system and the approximated time in which the system will remain functioning. Additional, either rotational or linear, movements of the material may apply.

g. Set the wire gauge and the number of turns per coil. This step will determine the output resistance. In addition, use Ohm's law to determine the desired output voltage, previously set in step one.

h. Make the coils with its respective insulating method to avoid losses by short circuit.

i. Assemble the system by setting up the magnet in a desired distance from the coils without colliding. The distance will determine how efficient the system will be. For instance, if it is too close, then there will be a great magnetic interaction so that a great time-varying magnetic field is assured for a better mechanical energy conversion.

j. Measure the generated output power values.

k. If any, check for necessary improvements and start over.

Meanwhile, this diagram represents the predecessor of the design, before making accurate final measures. Of course, either simulating programs or CAD software designing is necessary to project the functionality of the system. **Figure 7** explains the feasibility to design electric machines according to the desired application. Equally, the electromagnetic transduction and thermodynamic laws apply for any power generator development.

Likewise, it is important to define the number of desired poles since this property determines the effectiveness of the designed system. Thus, a system with lesser number of poles can generate energy at greater velocities than if the system is structured with a greater number of poles since it would carry out to slow down the velocity caused by the Lorentz forces and low velocities would take effect. Moreover, the addition of poles in the design yields to change the magnet's shape with the aim to design freely a power generation system with the highest magnetic values, and consequently, a very efficient power source.

The aim of using any electric machine is to increase efficiency to generate much more energy to satisfy the increasing energy demand. Of course, the necessity to manufacture variable efficiency power generators can result in overcoming the power generation with the current rotational velocities permitted. In consequence, the results are the reliability and durability that these electric machines can provide, by designing a power generator since the beginning. Hence, there are a few options to regard:

- The development of a model according to the output power needed.

- Additionally, the sum of energy systems is permitted to reach the targeted power.

- The possibility to harness rotatory mechanical energy by connecting several generators serially.

- The option to connect different number of phases according to the power consuming application.
- Voltage fluctuations and rotational velocities are allowed but with the aim to harvest energy.
- A contribution to renewable energies by the creation of an environmentally friendly option to power supply. Furthermore, to generate exergy in places where it is difficult to supply energy.
- The conversion of mechanical energy beyond velocity ranges of current generators.
- The development of customized generators by implementing the development diagram to be assembled anywhere and by anyone.
- The result of creating an optional power source for real life situations.
- The possibility of low lifetime maintenance and costs.

These points must be accomplished with the aim to target energy necessities around the world. The use of renewable energies will cause an effect on reducing pollution eventually. Hereafter, the development of electric machinery instead of combustion generators is a step forward to evolve more efficient technologies.

5.1 Equivalent circuit

The design of electric machinery according to the diagram in **Figure 7** has implicitly an equivalent circuit. The electrical characteristics of the implemented circuit are described by the designer. The equivalent circuit is the general representation of usual designs by implementing the previously addressed materials. Indeed, it is stated that the output voltage made from a design would be a sinusoidal wave (AC Voltage—$V_{\varnothing}$). Additionally, the permanent magnet is represented as a variability VAR, since in every time-varying magnetic field, there exist an electric current [39]. Moreover, the separation between the iron screw and the ferrite magnet is represented by the air gap, necessarily for both, to prevent the magnet from colliding and determine the efficiency of the system (**Figure 8**).

As the harvested energy is generated by the movement of a varying magnetic field, the electrical frequency is taken as

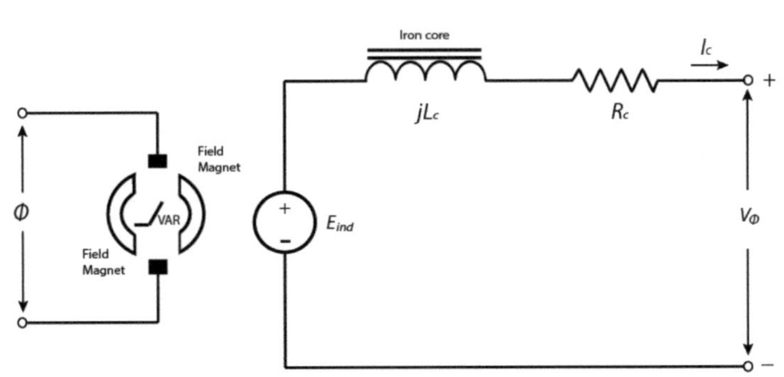

Figure 8.
Equivalent circuit of a modeled generator.

$$f_e = \frac{p \cdot \omega}{2\pi} \tag{1}$$

where p is the number of poles and ω is the angular velocity in radians. Of course, a single conversion can be made in Eq. (1) to calculate it with a linear velocity [40].

By knowing the N_c turns in a coil, where it is placed around a magnetic field ∅, the induced energy in the coil will be

$$e_c = N_c \varnothing \omega \cos \omega t \tag{2}$$

Accordingly, the peak voltage can be calculated as

$$V_{peak} = N_c \varnothing \omega = 2\pi N_c \varnothing f_e \tag{3}$$

Thus, the induced energy in RMS is stated as

$$E_{ind} = \frac{2\pi}{\sqrt{2}} N_c \varnothing f_e = \sqrt{2}\pi N_c \varnothing f_e \tag{4}$$

Therefore, as the stator is considered as the armature structure in the designs, the internally generated voltage in a single phase E_{ind} (from one to n-number of serially connected coils) is not usually the output voltage $V_{\varnothing}$, due to the resistance of the materials and the distortion caused by the air gap magnetic field [40] and in this case, the space between the permanent magnet and the armature (**Figure 8**).

According to **Figure 6**, the load current I_c generates a magnetic field B in the coil that will generate a voltage reaction named E_{coil}. Furthermore, a voltage reaction E_{ind} is induced by the time-varying magnetic field every time there exists an alignment of the magnetic domains [41].

Hence, the output voltage per phase is stated by

$$V_{\varnothing} = E_{ind} + E_{coil} \tag{5}$$

Thereby, the induced energy in the circuit can be modeled by

$$E_{coil} = -jL_c I_c \tag{6}$$

where $V_{\varnothing}$ represents the voltage generated. Because of Eq. (5) and Eq. (6), the output voltage is

$$V_{\varnothing} = E_{ind} - jL_c I_c \tag{7}$$

Thus, the equivalent current regarding the output resistance of the connected coils is

$$V_{\varnothing} = E_{ind} - jL_c I_c - R_c I_c \tag{8}$$

Therefore, the output voltage per phase results to be

$$V_{\varnothing} = \sqrt{2}\pi N_c \varnothing f_e - I_c (jL_c + R_c) \tag{9}$$

In contrast, it is assumed that in each interaction, a time-varying magnetic field crosses the coils, so that the magnetic flux is uniform and the emf is distributed evenly throughout each coil.

The efficiency of any electric machine can be immediately perceived on the hardness to move the magnets in accordance with the coils, for instance, the harder, the more efficient, and vice versa. Although the electromagnetic generators are a simple application of Faraday's law, they are a useful tool to satisfy present needs for energy source applications for dynamic systems without pollution.

One great advantage of these electromagnetic systems is that they are frictionless, so there is no weathering of the pieces, excepting the friction on the axis where rotation occurs and Lorentz damping forces. Thus, the cost-benefit increases, thanks to lower lifetime maintenance expense and long endurance. This is the future drive technology of the next years to come for a better-quality energy and more availability every time the energy is required as a tool to the energy shortage. Therefore, these designs are the benchmark for configurable generators, making a step forward in the evolutions of power system generators immersed in novel power cycles.

Finally, it is important to apply an exergy analysis at the process and component level. In consequence, it allows identifying, locating, and quantifying the main irreversible causes of thermodynamics of a system or process, through the study of exergy destruction and efficiency. Moreover, as the exergy is the available part of the energy used to produce useful work, it represents a powerful tool to determine potential improvements and optimization of processes with electromagnetic transduction application, as well as the mitigation of environmental concerns which results in a measure of the imbalance with the environment.

6. Conclusions

An exergy balance is the combination of the energy balance and entropy, since they are derived from the first and second principles of thermodynamics. However, it is an additional tool to make efficient the second law of thermodynamics.

Meanwhile, as an alternative to the increasing principle of entropy, the second law can state that the only processes that an isolated system can experience are those in which the exergy of the system decreases.

The balance of exergy is a very useful method of analysis when assessing the energy performance of a system by giving a broader vision than a thermal performance. Further, it allows evaluating the losses of energy in a process, the energy that will be utilized from outgoing flows in an open system and the advantages of regenerative methods in machines that get heated up easily. If a thermal machine is not performing with additional procedures to increase efficiency, then the option to replace it by a new technology that accomplishes energy necessities will take effect.

Nonetheless, since electromagnetic transduction has demonstrated to have a better performance than thermal machines, renewable energies and effcient power-consuming systems may replace them in its whole. For instance, ferromagnetic materials have a relative permeability of several hundred thousands, and they are strongly attracted by magnetic fields. This is the reason to employ ferromagnetic materials instead of paramagnetic and diamagnetic materials, causing the effect to generate time-varying magnetic fields, according to both, the magnetic flux and electric fields characteristics of ceramic ferrites. However, it has been outlined that neodymium magnets are magnetically three hundred times denser and stronger. This effect is possible, thanks to the relative permeability and magnetic susceptibility properties of these rare-earth magnets.

In recent years, the necessity to develop and to design novel renewable energy systems has been on the rise due to environmental impact by fossil fuels and the eventual depletion of the reserves of this hydrocarbon. In the meantime, these materials are more adaptable for electric field distribution. Further, by harnessing the mechanical energy, it is possible to power supply in a practical form several real-life applications. The required material is by implementing either coils or shunts and magnets in the system designing.

In contrast, the resulting output power depends crucially on the design, the covered area among the magnet and coils, the lineal or angular velocity induced, the relative permeability of the implemented materials, and the time-varying magnetic field generated, as well as the magnetic flux density and magnetic field strength of the magnets. The designs can be a part of a larger mechanical conversion energy system as they require an external mechanical force to harvest the energy from wind, tidal, or hydraulic energy systems. Furthermore, the characteristics of the implemented materials may either increase or decrease the performance of the power generation systems.

The electromagnetic transduction represents a modern background among renewable energy power systems, exergy analysis, and the electromagnetic engineering concepts to ensure long endurance, instant energy generation, efficiency, performance, and optimization of the energy. The applications would primarily be focused on low power consumption but projected to be for high power applications with the aim to enable more effective systems than current technologies over the 30 to 50 years to come.

Conflict of interest

The author declares no conflicts of interest.

Appendices and nomenclature

B	magnetic field
BMS	battery management system
E	electric field
emf	electromotive force
G	Gauss—magnetic flux density
H	magnetic field strength
M_{sat}	magnetization saturation
NdFeB	neodymium magnet
Oe	Oersted—magnetic strength
$Ø_m$	time-varying magnetic field
rpm	revolution per minute
T	tesla
μ	air permeability
μ_0	relative permeability
V	output voltage
W	Watt
W/kg	specific power
W/m^3	power density
X_m	magnetic susceptibility

Author details

Eduardo Torres-Sánchez
Tecnológico de Monterrey, Mexico City, Mexico

*Address all correspondence to: e89.torres10@gmail.com

References

[1] Kanoglu M, Dincer I, Rosen MA. Understanding energy and exergy efficiencies for improved energy management in power plants. Energy Policy. 2007;**35**(7):3967-3978

[2] Harb A. Energy harvesting: State-of-the-art. Renewable Energy. 2011; **36**(10):2641-2654

[3] Rant Z. Exergie, ein neues Wort fur 'Technische Arbeitsfaehigkeit' (exergy, a new word for technical availability). Forschung auf dem Gebiet des Ingenieurwesens A. 1956;**22**(1):36-37

[4] Ahern JE. Exergy Method of Energy Systems Analysis. New York: Wiley; 1980

[5] Szargut J, Morris DR, Steward FR. Exergy Analysis of Thermal, Chemical, and Metallurgical Processes. New York: Hemisphere Publ. Corp.; 1987

[6] Utlu Z, Hepbasli AJR. A review and assessment of the energy utilization efficiency in the Turkish industrial sector using energy and exergy analysis method. Renew Sustain Energy. 2007; **11**(7):1438-1459

[7] Dincer I, Rosen MA. Exergy: Energy, Environment and Sustainable Development. Kidlington: Newnes; 2012

[8] Burger S, Chaves-Ávila JP, Batlle C, Pérez-Arriaga IJ. A review of the value of aggregators in electricity systems. Renewable and Sustainable Energy Reviews. 2017;**77**:395-405

[9] Asif M, Muneer T. Energy supply, its demand and security issues for developed and emerging economies. Renewable and Sustainable Energy Reviews. 2007;**11**(7):1388-1413

[10] Segall P. Earthquakes triggered by fluid extraction. Geology. 1989;**17**(10): 942-946

[11] Meng Q, Ashby S. Distance: A critical aspect for environmental impact assessment of hydraulic fracking. The Extractive Industries and Society. 2014; **1**(2):124-126

[12] Upton GB Jr, Snyder BF. Funding renewable energy: An analysis of renewable portfolio standards. Energy Economics. 2017;**66**:205-216

[13] Challa VR, Prasad M, Fisher FT. A coupled piezoelectric–electromagnetic energy harvesting technique for achieving increased power output through damping matching. Smart Materials and Structures. 2009;**18**(9): 095029

[14] Blaabjerg F, Chen Z, Kjaer SB. Power electronics as efficient interface in dispersed power generation systems. IEEE Transactions on Power Electronics. 2004;**19**(5):1184-1194

[15] Pérez-Lombard L, Ortiz J, Pout CJE. A review on buildings energy consumption information. Energy and Buildings. 2008;**40**(3):394-398

[16] Belhora F et al. Mechano-electrical conversion for harvesting energy with hybridization of electrostrictive polymers and electrets. Sensors and Actuators A: Physical. 2013;**201**:58-65

[17] Amanor-Boadu J, Abouzied M, Carreon-Bautista S, Ribeiro R, Liu X, Sanchez-Sinencio E, editors. A switched mode Li-ion battery charger with multiple energy harvesting systems simultaneously used as input sources. In: 2014 IEEE 57th International Midwest Symposium on Circuits and Systems (MWSCAS). Texas: IEEE; 2014. pp. 330-333

[18] El-Azab A, Garnich M, Kapoor A. Modeling of the electromagnetic forming of sheet metals: State-of-the-art and future needs. Journal of Materials

Processing Technology. 2003;**142**(3): 744-754

[19] Kotas TJ, Flow F. Exergy concepts for thermal plant: First of two papers on exergy techniques in thermal plant analysis. International Journal of Heat and Fluid Flow. 1980;**2**(3):105-114

[20] Shukuya M, Hammache AJVT. Introduction to the Concept of Exergy-for a Better Understanding of Low-Temperature-Heating and High-Temperature-Cooling Systems. Finland: VTT Tiedotteita; 2002

[21] Liu H, You L. Characteristics and applications of the cold heat exergy of liquefied natural gas. Energy Conversion and Management. 1999; **40**(14):1515-1525

[22] Fahie J. Magnetism, electricity and electromagnetism up to the time of the crowning work of Michael Faraday in 1831: A retrospect. Electrical Engineers, Journal of the Institution of. 1931; **69**(419):1331-1357

[23] Wilczek F. Magnetic flux, angular momentum, and statistics. Physical Review Letters. 1982;**48**(17):1144

[24] Coey JM. Magnetism and Magnetic Materials. New York: Cambridge University Press; 2010

[25] Tipler PA, Mosca G. Physics for Scientists and Engineers. 6th ed. New York: W.H.Freeman & Co Ltd; 2007

[26] Takezawa H, Hirakawa N, Mohri N. Surface magnetic flux density patterning in EDM of permanent magnets. Procedia CIRP. 2016;**42**:668-672

[27] Landau L, Lifshitz E. On the theory of the dispersion of magnetic permeability in ferromagnetic bodies. Physikalische Zeitschrift der Sowjetunion. 1935;**8**(153):101-114

[28] Bleaney BI, Bleaney BI, Bleaney B. Electricity and Magnetism. Vol. 2. Oxford: Oxford University Press; 2013

[29] Darwin CG. The diamagnetism of the free electron. In: Mathematical Proceedings of the Cambridge Philosophical Society. Cambridge: Cambridge University Press; 1931. pp. 86-90

[30] Stoner EC. Collective electron specific heat and spin paramagnetism in metals. Proceedings of the Royal Society of London. Series A, Mathematical and Physical Sciences. 1936;**154**(883): 656-678

[31] Bozorth RM. Ferromagnetism. In: Bozorth RM, editor. Ferromagnetism. New York: Wiley-IEEE Press; 1993. p. 992. ISBN 0-7803-1032-2

[32] Della Torre E. Magnetic Hysteresis. New York: Wiley; 2000

[33] Sugimoto M. The past, present, and future of ferrites. Journal of the American Ceramic Society. 1999;**82**(2): 269-280

[34] Benz M, Martin DL. Cobalt-Samarium permanent magnets prepared by liquid phase sintering. Applied Physics Letters. 1970;**17**(4):176-177

[35] Croat J. Current status and future outlook for bonded neodymium permanent magnets. Journal of Applied Physics. 1997;**81**(8):4804-4809

[36] Sagawa M, Fujimura S, Togawa N, Yamamoto H, Matsuura Y. New material for permanent magnets on a base of Nd and Fe. Journal of Applied Physics. 1984;**55**(6):2083-2087

[37] Coey JMD. Rare-Earth Iron Permanent Magnets. Oxford: Oxford University Press; 1996

[38] Lee JH, Hyun DS. Hysteresis analysis for the permanent magnet assisted synchronous reluctance motor by coupled fem and preisach modelling. IEEE Transactions on Magnetics. 1999; **35**(3):1203-1206

[39] Boldea I. Synchronous Generators. 2nd ed. Boca Raton: CRC Press, Taylor & Francis Group; 2015

[40] Kron G. Equivalent Circuits of Electric Machinery. 1st ed. Hoboken: John Wiley & Sons; 1951

[41] Krause PC, Wasynczuk O, Sudhoff SD, Pekarek S. Analysis of Electric Machinery and Drive Systems. Hoboken: John Wiley & Sons; 2013